Molecular Mechanisms in Materials

Molecular Mechanisms in Materials

Insights from Atomistic Modeling and Simulation

Sidney Yip

The MIT Press
Cambridge, Massachusetts
London, England

The MIT Press would like to thank the anonymous peer reviewers who provided comments on drafts of this book. The generous work of academic experts is essential for establishing the authority and quality of our publications. We acknowledge with gratitude the contributions of these otherwise uncredited readers.

This book was set in Stone Serif and Stone Sans by Westchester Publishing Services. Printed and bound in the United States of America.

Library of Congress Cataloging-in-Publication Data

Names: Yip, Sidney, author.
Title: Molecular mechanisms in materials : insights from atomistic modeling and simulation / Sidney Yip.
Description: Cambridge, Massachusetts : The MIT Press, [2023] | Includes bibliographical references and index.
Identifiers: LCCN 2022052640 (print) | LCCN 2022052641 (ebook) | ISBN 9780262048132 (paperback) | ISBN 9780262374958 (epub) | ISBN 9780262374965 (pdf)
Subjects: LCSH: Materials science—Computer simulation.
Classification: LCC TA404.23 .Y57 2023 (print) | LCC TA404.23 (ebook) | DDC 620.1/10113—dc23/eng/20230516
LC record available at https://lccn.loc.gov/2022052640
LC ebook record available at https://lccn.loc.gov/2022052641

10 9 8 7 6 5 4 3 2 1

To my mentors, former students,
and collaborators,
a journey of enlightenment

Contents

Preface

The reader may regard this book as a self-organized art exhibit where the author is the artist-curator. The exhibit is a collection of essays, each being a story of materials research with emphasis on physical understanding. While the scientific findings discussed have already appeared in publications, the collection as a whole has not been digested from the standpoint of a retrospective. It is my hope this undertaking is useful to newcomers—students early in their research, or scientists from related fields looking for an entry into the subject domain herein.

After coediting the second edition of the *Handbook of Materials Modeling* (2020) with Wanda Andreoni, I found myself facing three options: do nothing more, do something different, or do one more small follow-up by myself. This work is the decision I made and ended up pursuing before and during the COVID-19 pandemic. I have learned much in ways I had not anticipated. Looking back, I am glad I took this path. I feel ever more confident there are rewards awaiting those who choose to continue pushing the frontiers in understanding materials phenomena.

This monograph is about the dynamical mechanisms in simple matter and their characterization through modeling and simulation at the molecular level. It attempts to present a unifying view of several basic types of materials behavior: liquid fluctuations, crystal melting, strength and plasticity, viscous transport, and amorphous rheology. As indicated in the table of contents, each essay addresses a specific, standalone topic of materials phenomena, while at the same time being mindful of relevance to the greater enterprise of materials science and technology.

In contrast to the traditional monographs, we adopt an essay-style of exposition in the interest of readability. Technical details are mostly relegated to the references cited. Additional reading is provided as suggestions for further study. An overall introduction and a concluding perspective appear as the prologue and epilogue, respectively. Figure 0.1 shows the organizational relationship between the five parts and the fifteen essays.

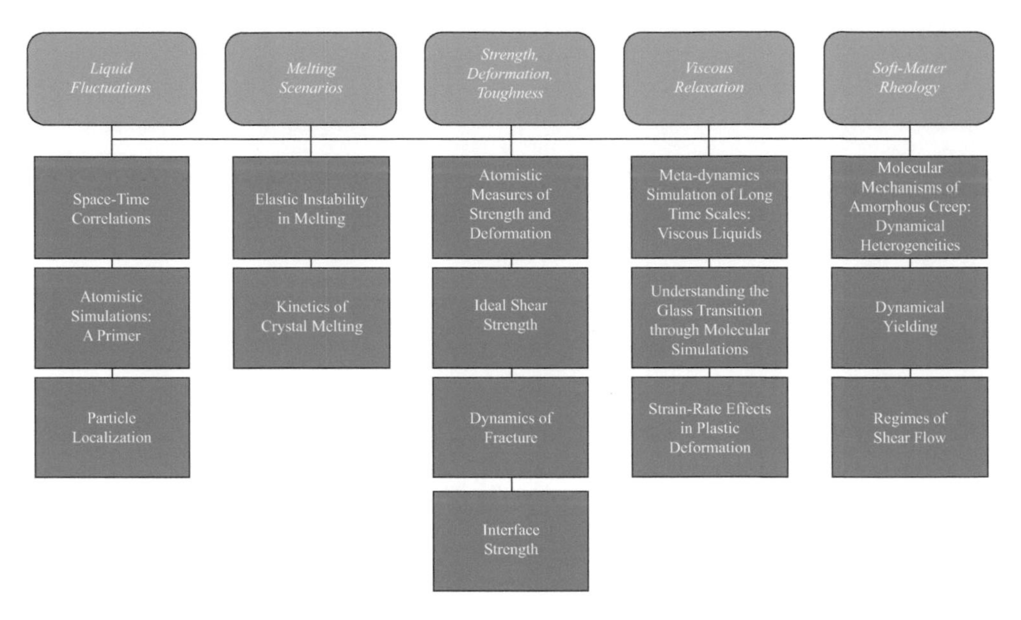

Figure 0.1
Flow chart of the essay collection corresponding to the table of contents.

This work is mostly based on collaborative research in which I have been involved over a span of fifty plus years. I am grateful to colleagues who were mentors, former students, and collaborators on various occasions. They were influential in what I have learned and the perspectives I have held over time. Although this book is dedicated to them collectively, I take this opportunity to recall enlightening discussions with the following: Farid Abraham, B. J. Alder*, Ali Argon*, Martin Bazant, Jean-Pierre Boon, Vasily Bulatov, Gerd Ceder, Sow-Hsin Chen*, Karin Dahmen, Shankar Das, James W. Dufty, Maurice de Koning, Emanuela Del Gado, P. A. Egelstaff*, Joshua Fujiwara, Wolfgang Goetze*, Jeff Grossman, Hamlin Jennings*, John Joannopulos, Paul Ho, Efthimios Kaxiras, Akihiro Kushima, James S. Langer, Eberhard Leutheusser, Francesco Mallemace, Dimitrios Maroudas, Paul C. Martin*, Gene Mazenko, Mark Nelkin, Shigenobu Ogata, Richard K. Osborn*, Anees Rahman*, Alf Sjolander*, Subra Suresh, Sam Trickey, John Ullo*, Franz Ulm, Henri Van Damme, Krystyn Van Vliet, Priya Vashihta, C.-Z. Wang, Dieter Wolf, Dongsheng Xu, Bilge Yildiz, and Paul Zweifel*.

The support of my home department has been invaluable during the writing process, Paula Cornelio on figure preparation, and Kristi Stone on all aspects of editorial assistance and word-processing, without which this book simply could not be produced. I have also leaned on colleagues Ju Li and Mike Short for help on computer issues.

Having been at the Institute for one's entire career, my desire to publish with the MIT Press is an easy decision. I thank Jermey Matthews and Haley Biermann for guiding this project to fruition.

**in memoriam*

Sidney Yip

Professor Emeritus
Department of Nuclear Science and Engineering
Department of Materials Science and Engineering
Massachusetts Institute of Technology

Cambridge, MA and Santa Barbara, CA
June 2022

Prologue: Computational Materials

The materials world is evolving rapidly driven by critical societal needs worldwide. Science innovations and technology breakthroughs allow the materials community to functionalize assemblies of atoms and molecules unimagined only a short time ago. Concurrently, high-performance computing brings artificial intelligence, machine learning, and decision-making capabilities into the forefront of all human activities. In this environment, computational materials thrives as a multidisciplinary enterprise, offering a wide range of opportunities and challenges in research and education.

Introduction

Underlying the study of molecular mechanisms in materials—the subject of this book—is the hypothesis that matter is composed of atoms and molecules, and all materials properties and behavior arise from the interactions among these constituent particles. The interactions will vary with the particular forms of the matter along with the environment in which it is being considered. Atoms in a liquid are expected to interact with each other differently from atoms in a crystalline lattice. How the constituents affect each other and the environment in which the interactions take place—the mechanisms—clearly will be very consequential.

Materials of all varieties are indispensable to human existence, for security to societal well-being, technological innovations and sustainability. Basic to the enterprise is our understanding and ability to manipulate materials, from natural resources to highly sophisticated synthetic devices, based on the concept of "mechanisms." The study of materials mechanism is tantamount to understanding how nature works.

Multiscale Modeling and Simulation

Problems in the fundamental description of matter, which previously were regarded as intractable, are now amenable to simulation and analysis. The ab initio calculation of solid-state properties using electronic-structure methods and the direct estimation of free energies based on statistical mechanical formulations are two such examples. Because materials modeling draws from all the disciplines in physical science and engineering, it can have a very broad impact in cross fertilization between traditionally different communities. About two decades ago, it was suggested that computational materials, the intersection between computational science and materials research, was just as vital as the longer established disciplines of computational physics or computational chemistry (Yip 2003). Each provides a robust framework for focused scientific studies and exchanges, from the introduction of new university curricula to the formation of centers for collaborative research among academia, corporate, and government laboratories. The appeal to all members of this new community is the integrated advancement of fundamental understanding with technology impacts for societal benefits (SBES 2006).

The organizing principle of computational materials is multiscale modeling and simulation, where conceptual models and simulation techniques are linked across the micro-to-macro length and timescales for the purpose of controlling the outcome of specific materials processes. Invariably these phenomena are highly nonlinear, inhomogeneous, or nonequilibrium. In this paradigm, electronic structure would be treated by quantum mechanical calculations, atomistic processes by molecular dynamics or Monte Carlo simulations, mesoscale microstructure evolution by methods such as finite element, dislocation dynamics, or kinetic Monte Carlo, and continuum behavior by field equations central to continuum elasticity and computational fluid dynamics. By combining these different methods, one can deal with complex problems in an integral manner that is not possible when the methods are used individually.

Modeling is the physicalization of a concept,
Simulation is its computational realization.

This is a highly simplified way to describe the two processes that give computational materials its unique character. Since there appears to be no consensus on what each term means by itself and in what sense one complements the other, we suggest here an all-purpose definition that is brief and general. By concept, we have in mind an idea, an idealization, or a picture of a system's behavior. By computational realization, we mean enacting the concept, making it come alive by means of simulation. In these definitions, the two are clearly complementary to each other.

Characteristic length/timescales Materials mechanisms have significant manifestations on more than one level of length and timescale. Generally, there are four distinct length (and, correspondingly, time) scales where materials behavior are typically studied. As illustrated in figure P.1, they may be referred to as electronic structure, atomistic,

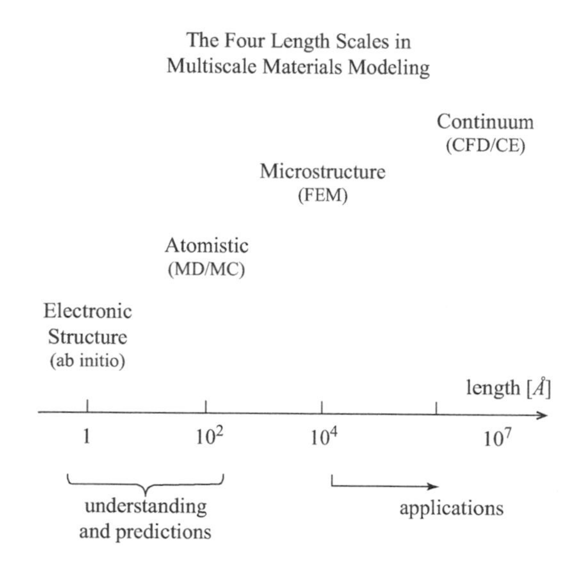

Figure P.1
Length scales in materials modeling. Many traditional materials applications take place on the micron scale and higher, while scientific understanding and predictive ability tend to lie mostly at microscopic levels. Reprinted by permission from Springer Nature: Yip (2006).

microstructure, and continuum (Yip 2005). Imagine a crystalline solid where the smallest length scale of interest is about a few angstroms (10^{-8} cm). On this scale, one deals directly with the electrons in the system that are governed by the Schrödinger equation in quantum mechanics. Because the techniques for solving the Schrödinger equation are computationally intensive, they are applied only to small simulation systems, usually a few thousand atoms. These calculations are the most rigorous from the theoretical standpoint. They are valuable for developing and testing the more approximate but computationally tractable descriptions.

The scale at the next level, spanning from tens of thousands to about a million or more atoms, is called atomistic. Here, discrete particle simulation techniques, principally molecular dynamics and Monte Carlo, are well developed, requiring the specification of an empirical classical interatomic potential function with parameters fitted to experimental data or electronic-structure calculations. The advantage of atomistic simulation is the molecular interaction mechanisms need not be specified a priori. On the other hand, ignoring the electrons means the results generally will not be as reliable as ab initio calculations.

Above the atomistic level, the relevant length scale is a micron (10^4 angstrom). Whether this level should be called microscale or mesoscale is somewhat arbitrary, the convention not being clearly established. The simulation technique commonly in use is the finite-element method. Because many useful properties of materials are governed by the microstructure of the system, this is perhaps the most relevant level for materials performance and design. However, the information required to carry out the calculations, for example, the stiffness matrix, or any material-specific physical parameter, has to be provided externally from either experiment or calculations at the atomistic or ab initio level. To a large extent, the same can be said for the continuum level, namely, the parameters needed to perform the calculations have to be supplied externally.

There are definite benefits when simulation techniques at different scales can be linked. Continuum or finite-element methods are often most practical for design calculations. They require parameters or properties which cannot be generated within the methods themselves. Also, they cannot provide the atomic-level insights needed for design. For these reasons, they need to be coupled to atomistic and ab initio methods. It is only when methods at different scales are effectively integrated that one can expect materials modeling to give fundamental insights as well as reliable predictions across the scales. The efficient bridging of the scales in figure P.1 is a continuing challenge in the evolution of computational materials (SBES 2006; Yip 2005).

Intellectual merits It would be appropriate to ask what are the intellectual merits of computational materials considered most significant to the global materials science and technology community. Three elements particularly stand out in the current era of molecular design: exceptional bandwidth, empiricism reduction, and visualization insights. Let us recall the principle of materials design is rooted in the correlation of molecular structure with physical properties to formulate predictive models of microstructure evolution. Through these relations, one can manipulate the underlying mechanisms to systematically arrive at improved designs. Structure-property correlation established by simulation can be superior to relying completely on experimental data because one has complete information on the evolving microstructure, as well as full control over the initial and boundary conditions.

Exceptional bandwidth The conceptual basis of materials modeling and simulation covers all of physical science without regard to what belongs to physics versus chemistry versus engineering. This means the scientific bandwidth of computational materials is as broad as all of the affiliated science and engineering.

Removing empiricism A virtue of multiscale modeling is that results pertaining to both model and simulation are conceptually and operationally quantifiable. This means that empirical assumptions can be systematically replaced by science-based descriptions. Quantifiability implies that any part of the modeling and simulation can be scrutinized and upgraded in a controlled manner, allowing a complex phenomenon to be probed part by part.

Visualization insights Simulation output are numerical data on the degrees of freedom characterizing the model. Their availability allows not only direct animation, but also visualization of analyzed properties which are not directly accessible to experimental observation. In microscopy, for example, one has structural information but usually without the energetics, whereas by simulation one can have both. The same may be said of deformation mechanisms and reaction pathways.

It has previously been noted that the foundation of computational materials is the conceptualization of a materials problem (modeling) combined with a computational solution of this problem (simulation) (Yip 2005). This coupled endeavor can be extended to many areas of science and technology. It practically allows any complex system to be analyzed with predictive capability by invoking the multiscale paradigm—linking unit-process models at lower length (or time) scales where fundamental principles have been established to calculations at the system level. It allows the understanding and visualization of cause-effect through simulations where initial and boundary conditions are prescribed specifically to gain insight. Furthermore, it can complement experiment and theory by providing the details that cannot be measured nor described through equations. When these conceptual advantages in modeling are coupled to continually advance in computing power through simulation, one has a vital and enduring scientific approach destined to play a central role in solving the formidable problems of our society. Yet, to translate these ideals into successful applications requires specific case studies. This book is an example of a collection of such materials research efforts.

Synergistic Science

Early recognitions of the promise of computational materials came from the US federal agencies of science and technology. In 2005, the President's Information Technology Advisory Committee stated in its report (PITAC 2005):

> Computational Science—the use of advanced computing capabilities to understand and solve complex problems—has become critical to scientific leadership, economic competitiveness, and national security.
>
> Computational science is now indispensable to the solution of complex problems in every sector . . . using computational models to capture and analyze unprecedented amounts of experimental and observational data, and to address problems previously deemed intractable.

This statement sets forth an overarching motivation for alliance among academia, industry, and government agencies.

At about the same time, the US National Science Foundation issued a report, *Simulation-Based Engineering Science*, stating "The National Science Foundation believes that computer simulation is developing into one of the most important areas of intellectual activity and research in modern history" (SBES 2006).

In these high-level assessments, one cannot expect too much specificity in explicitly identifying the complex problems that modeling and simulation can be expected to yield in new understanding and solution. Thus, it was left to subsequent topical conferences, workshops, and joint programs to explore and assess specific topics as part of the community building process. A few will be mentioned here as illustrative examples.

Nuclear Materials Challenges, DOE Workshop on Basic Research Needs for Advanced Nuclear Energy Systems, July 2006 An overall theme of this workshop, which brought together the computational materials research communities in academia and the various national laboratories in the US, was simply stated as:

> Understand and control chemical and physical phenomena in multi-component systems—femto seconds to millennia, temperatures to 1000°C, and radiation doses to hundreds of displacements per atom.

This succinct statement exemplifies what one could mean by materials behavior in the extreme. It can be taken to be a grand challenge in computational materials that is directly relevant to society. Another example in the same context is the understanding and control of nuclear waste, where timescales of millennia are of primary concern. In this case, laboratory experiments on the scale of months are of questionable value; one can only resort to statistical theories of transport and high-performance simulations capable of sampling rare events.

Accelerated Strategic Computing Initiative (ASCI 2005–2015) In support of the Nuclear Test Ban Treaty, the US Department of Energy decided to stop all nuclear weapons testing for component reliability and maintenance and launched a decade-long program aiming to directly simulate the performance of a nuclear warhead entirely on the computer. The initiative involved the development of more powerful supercomputers, along with concurrent software advancement. In the materials arena, a major goal was to determine the mechanical behavior (yield surface) of bcc metals (surrogate for plutonium) using multiscale modeling and simulation (Larzelere 2009; ASC 2012).

CASL (Consortium for the Advanced Simulation of Light Water Reactors) This was another decade-long DOE Energy Innovation Hub alliance between the DOE laboratories, academia, and industrial vendors. In contrast to ASCI, this program focused on innovations to improve the safety, reliability, and performance of the light water reactor fleet in the US. Three industrial problems were specified as targets for research by the entire consortium: grid lock fretting (fatigue of the fuel rod alignment plate), CRUD (buildup of ion deposits from the coolant), and fuel-clad interaction (hydrogen transport through the Zircaloy cladding) (CASL 2020).

Handbook of Materials Modeling and Comprehensive Nuclear Materials There are two major reference works serving the growing community of computational materials. Positioned at the intersection of computational science with materials science and technology, the second edition of the *Handbook of Materials Modeling* consists of two mutually dependent sets—*Methods: Theory and Models* (MTM), and *Applications: Current and Emerging* (ACE)—each in three volumes (Andreoni and Yip 2020b; Andreoni and Yip 2020a). *Comprehensive Nuclear Materials*, on the other hand, is focused on materials in nuclear power generation. The second edition, in seven volumes, has recently been released (Konings and Stoller 2020).

Multiscale Materials Modeling (MMM) International Conferences This series started in 2002 and has continued every two years, with steadily growing attendance reaching 600–700 participants. The venue distribution is shown below.

The MMM Conferences

I	London (2002)
II	UCLA (2004)
III	Freiburg (2006)
IV	Tallahassee (2008)
V	Freiburg (2010)
VI	Singapore (2012)
VII	Berkeley (2014)
VIII	Dijon (2016)
IX	Osaka (2018)

The next MMM meeting is scheduled for October 2022 in Baltimore, Maryland. If one were to identify one community that is closely aligned with the activities of computational materials, it would be the combination of this community with another conference series called *Psik*, which could be regarded as the counterpart focused on first-principles modeling and simulation and based mostly in Europe.

Mesoscale Science Frontier

Around the period of 2010–2015, DOE decided to focus particular attention on research at the length scale that is intermediate between the atomistic and molecular (microscale) and the continuum (macroscale), to be denoted loosely as the *mesoscale*. For a few years, deeper understanding of how to connect the fundamental understanding of mechanisms at the microscale with the materials technology applications at the macroscale attracted widespread attention in the community, resulting in the idea of a mesoscale science research frontier (Crabtree 2012; Yip and Short 2013).

In 2012, the DOE Office of Science initiated a dialogue with the science community through a series of town hall meetings on identifying new science frontiers at the mesoscale. A website was created to solicit community input. A report, *From Quanta to the Continuum*, was released on how strategic materials research could impact the science and technology community at large. From these discussions, it seems that "mesoscale science" should be regarded as an *open* concept, the principles of which depend on the application context. In other words, the mesoscale science frontier (MSS) can be characterized in different ways. In an early assessment of what is mesoscale, the organizing principle was taken to be the energy landscape description of transition states, and in another formulation, it was self-organization with nonlinear feedback, and still from a third perspective, it could be frustration or localization effects leading to jamming phenomena of interest in statistical or biophysics communities. If one could call this a top-down approach, then a bottom-up approach could focus on discrete-particle simulation and extract local spatial and temporal correlations to identify activated processes, dynamically heterogeneous environments, and intermittent rare events.

To be more explicit, materials problems have been put forth as mesoscale challenges, thus helping to illustrate the broad class of phenomena that one could study as opportunities to advance computational materials. The present collection of essays, individually and collectively, may be regarded from such a perspective. The spirit of MSS can be expressed graphically as a research frontier (boundary) between science on the one side and technology on the other, as indicated in figure P.2.

Essay Overviews

In choosing the fifteen essays presented in this work, our overall objective is physical understanding of molecular mechanisms in materials derived with the use of molecular simulation. While each essay deals with a particular mechanism, broader insights and implications are possible by virtue of overlaps with other essays, particularly the companion ones in the same topical group. The fifteen essays fall into five basic types of materials phenomena: fluctuations in simple fluids, crystal melting, plasticity and fracture, glassy relaxations, and amorphous rheology. The topical contents of each essay are briefly indicated in table P.1.

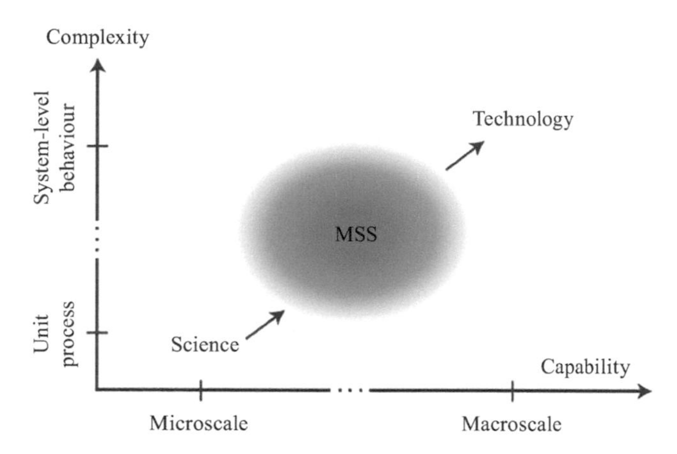

Figure P.2
Complexity-capability map showing the gap between science research at the microscale and technology applications at the macroscale. Bridging this gap requires the integration and collaboration between scientists and technologists in the community, sharing common goals of overarching societal benefits. Figure from (Yip and Short 2013).

Table P.1
Concepts and phenomena addressed in this book (listed by essay).

1	Space-time correlations, binary collisions, kinetic theory, hydrodynamics, memory function
2	MD (Newtonian dynamics) and MC (stochastic) simulations, unique features
3	Mode-coupling ideas, nonlinear feedback, particle localization in diffusion
4	Elasticity melting criteria, MD verification, instability triggered transitions
5	Free-energy based melting in Si, role of surface and defects
6	Plasticity response to stress activation, dislocation nucleation and mobility
7	Shear strength, deformation in metals and ceramics, charge-density effects
8	Reaction pathway simulation, crack tip nucleation-extension, brittle-ductile behavior
9	Crystal grain size effects on strength, insights from simulation
10	Long time-scale simulations, viscous behavior of supercooled liquids
11	Molecular nature of vitrification, understanding fragile and strong viscosity
12	Rate-dependent defect nucleation in crystals and glasses
13	Strain deformation, thermal activation versus stress localization, creep mechanism map
14	Yielding, stress-relaxation avalanches, shear band nucleation
15	Strain rate–modulated shear flow regimes, dynamical heterogeneity, potential-energy landscape

In the epilogue, we will return to a companion table to summarize the mechanism relevance of the essays for an outlook on future directions where computational materials are likely to thrive.

References

ASC (Accelerated Strategic Computing). https://asc.llnl.gov/role-asc-stockpile-stewardship; https://www.sandia.gov/asc/; https://www.lanl.gov/projects/advanced-simulation-computing/

Andreoni, W., and S. Yip, eds. 2020a. *Handbook of Materials Modeling, Applications: Current and Emerging (ACE)*, 2nd ed. Switzerland: Springer Nature.

Andreoni, W., and S. Yip, eds. 2020b. *Handbook of Materials Modeling, Methods: Theory and Models (MTM)*, 2nd ed. Switzerland: Springer Nature.

CASL. 2020. *CASL Phase II Summary Report, September 2020*. Accessed from https://casl.gov/wp-content/uploads/2020/11/CASL_FINAL_REPORT_09.30.2020-002.pdf

Crabtree, G. W., and J. L. Sarrao. 2012. "Opportunities for mesoscale science." *MRS Bulletin* 37: 1079.

Konings, R., and R. E. Stoller, eds. 2020. *Comprehensive Nuclear Materials*, 2nd ed. Oxford: Elsevier.

Larzelere, A. R. 2009. *Delivering Insight, The History of Accelerated Strategic Computing Initiative (ASCI)*. Report to the Lawrence Livermore National Laboratory under subcontract B545072.

PITAC. 2005. "Computational Science: Ensuring America's Competitiveness." Accessed from https://apps.dtic.mil/sti/pdfs/ADA462840.pdf

SBES. 2006. "Simulation-Based Engineering Science: Revolutionizing Engineering Science through Simulation." Report of the National Science Foundation Blue Ribbon Panel on Simulation-based Engineering Science. Accessed from http://www.nsf.gov/pubs/reports/sbes_final_report.pdf

Yip, S. 2003. "Synergistic science." *Nature Materials* 2: 3.

Yip, S., ed. 2005. *Handbook of Materials Modeling*. Dordrecht: Springer.

Yip, S., and M. P. Short. 2013. "Multiscale materials modelling at the mesoscale." *Nature Materials* 12 (9): 774.

Further Reading

US Department of Energy. 2007. *Delivering Insight—The History of the Accelerated Strategic Computing Initiative*. Accessed from https://asc.llnl.gov/file-download/download/public/2326. This report is rich and comprehensive in impact. It is recommended for the serious reader as a sustained effort in synergistic activity at the national level. The efforts are continuing under the auspices of three national laboratories.

Laughlin, R. B., D. Pines, J. Schmalian, B. P. Stojkovic, and P. Wolynes. 2000. "The middle way." *Proceedings of the National Academy of Sciences of the United States of America* 97: 33. A provoking commentary on the mesoscopic organization of soft, hard, and biological matter suggesting the possible existence of as-yet-undiscovered organizing principles that could bridge between fundamental principles at the molecular level and macroscale behavior at the continuum-system level.

Short, M. P., and S. Yip. 2015. "Materials aging at the mesoscale: kinetics of thermal, stress, radiation activations." *Current Opinion in Solid State and Materials Science* 19: 245.

Integrated Computational Materials Engineering (ICME) is a comprehensive program launched in 2011 by the US National Science and Technology Council to accelerate the discovery, design, development, and deployment of materials for societal benefits through the synergistic combination of computational and experimental capabilities involving all stakeholders, essentially in the same vision that is the motivation for this work; see the *2021 Materials Genome Initiative Strategic Plan* and the website of the 6th World Congress held in April 24–28, 2022, https://www.tms.org/icme2022.

I Liquid Fluctuations

1 Space-Time Correlations

The hypothesis underlying the study of molecular mechanisms, the subject of this book, is that matter is composed of atoms and molecules, and properties and behaviors of these systems therefore arise from the interactions among the constituent particles. The details of interactions will vary with the particular form of matter and the environment in which it is being considered. Atoms in a liquid can be expected to interact with each other differently from atoms in a crystalline lattice. Details of the interactions—the mechanisms—will matter. Moreover, if the system is not in equilibrium, then the external environment, be it thermal, mechanical, or chemical, becomes just as important in determining the system behavior. In other words, interaction mechanisms can have an extrinsic component for materials systems.

It was first realized in the community of nonequilibrium statistical mechanics that thermal fluctuations in dense fluids and liquids can be usefully studied through the concept of space-time correlation functions. These functions are not only well-defined conceptually, but they can also be calculated at varying levels of theoretical rigor and applied to interpret thermal neutron and laser scattering experiments. At about the same time, atomistic simulations were emerging to further stimulate the theoretical developments. Although applicable to all states of matter, the concept of space-time correlations focused initially mostly on liquids. In this opening essay, we introduce time correlation functions to provide a foundation on which theoretical studies and atomistic simulations (essay 2) can be combined toward understanding the dynamics of the liquid state. As a further noteworthy development, gaining insights into nonlinear dynamics (feedback mechanisms), we show in essay 3 how mode-coupling effects can lead to particle localization in a disordered, diffusive medium.

Introduction

Since liquids generally exist at finite temperature, the thermal fluctuations in these systems must play an important role in their properties and behavior. In the 1960s, there was a good deal of interest surrounding the study of liquid-state dynamics. There were two principal drivers. One was experimental in nature, the advent of neutron and laser inelastic scattering spectroscopy that could provide spectral data on liquids in different ranges of wavelengths and frequencies. The other was theoretical, the development of space-time correlation functions that describe the effects of interatomic interactions on the dynamical properties. The two developments were highly synergistic in that the new data becoming available served as motivation for theoretical explanation, and in return, theory provided incentives for further experiments. There was also a third driver that was just emerging: atomistic simulation, which could be used to either interpret experiments or validate theoretical calculations. The integration of theory-experiment-simulation, which is now standard practice in many fields of science and technology, essentially had its beginning during this period.

In this opening essay, we introduce the density time correlation function as the most important member of the family of time correlation functions, and show how this quantity can be studied by solving kinetic equations of transport at the molecular level and the equations of hydrodynamics at the level of continuum theory.

We start with basic definitions and simple calculations to introduce the family of space-time correlation functions. These quantities may be regarded as the frequency- and wavenumber-dependent generalizations of transport coefficients. Because they are defined in terms of equilibrium ensembles, it follows they are the most natural measures of the time-dependent thermal fluctuations that occur in a fluid at equilibrium.

Density Correlation Function

Since the density of a system is typically the most fundamental state variable and any change in the system is likely to affect the density, it is not surprising the density correlation function is basic to our understanding of thermal fluctuations in statistical mechanics.

For a system of N discrete particles in a volume V, the number density at the position $\underline{r}$ and time t can be written as

$$n(\underline{r},t) \equiv (1/\sqrt{N}) \sum_{\ell=1}^{N} \delta(\underline{r} - \underline{R}_\ell(t)) \tag{1.1}$$

where the thermal average $< n(\underline{r},t) >$ is just $\sqrt{N}/V$. The autocorrelation of density, which we call the density correlation function, is defined as

$$G(\underline{r}-\underline{r}', t-t') = V < \delta n(\underline{r},t) \delta n(\underline{r}',t') > \tag{1.2}$$

where the angular brackets $<>$ denote thermal average, and $\delta n(\underline{r},t) = n(\underline{r},t) - < n(\underline{r},t) >$ is the deviation in the density at position $\underline{r}$ and time t (Boon and Yip 1980). Writing out equation (1.2) more fully, one obtains

$$G(\underline{r},t)=\frac{1}{n}\sum_{\ell,\ell'=1}^{N}<\delta(\underline{R}_\ell(0))\delta(\underline{r}-\underline{R}_{\ell'}(t))>-n \quad (1.3)$$

We can take $\underline{R}_\ell(0)=0$ for a homogeneous system, then

$$G(\underline{r},t)=\sum_{\ell=1}^{N}<\delta(\underline{r}-\underline{R}_\ell(t))> \quad (1.4)$$

This definition shows that $G(\underline{r},t)$ has a simple physical meaning, namely, the average density at $\underline{r}$ at time t given that a particle was at the origin at $t=0$. Notice the correlation is both spatial and temporal.

There exists another density correlation function very similar to $G(\underline{r},t)$. It is the self-correlation or single-particle correlation function, usually denoted by the symbol $G_s(\underline{r},t)$. Imagine we can label a particular particle in the N-particle fluid, the so-called test particle. Other than the label, this particle behaves just as any other particle. Let this particle be labeled particle 1. Analogous to equation (1.1), the probability this tagged particle is at position $\underline{r}$ and time t is just

$$n_s(\underline{r},t)=\delta(\underline{r}-\underline{R}_1(t)) \quad (1.5)$$

and the corresponding density correlation function, called the van Hove self-correlation function, becomes

$$G_s(\underline{r},t)=V<\delta(\underline{R}_1(0))\delta(\underline{r}-\underline{R}_1(t))>-1/V \quad (1.6)$$

Similarly, the interpretation of Gs is the probability of finding a particle at position $\underline{r}$ at time t given it was at the origin at time $t=0$. Based on their physical meanings, it is clear that Gs describes the motion of a tagged particle as it moves through the fluid (diffusion), whereas G describes the motion of a particle relative to another particle at a different position and different time. Since the latter can be the same particle or a different one, it follows that $G(\underline{r},t)$ can be decomposed into the sum of $G_s(\underline{r},t)$ and another density correlation $G_d(\underline{r},t)$,

$$G(\underline{r},t)=G_s(\underline{r},t)+G_d(\underline{r},t) \quad (1.7)$$

where subscript d denotes *distinct*, referring to the particle at the origin at $t=0$ relative to the particle at position $\underline{r}$ and time t. It can be expected that G_d describes the cooperative or collective motions in the fluid. Figure 1.1 illustrates the two components of G, with G_s being the contribution from the particle initially at the origin (closed circle), and G_d the contribution from all the other particles (open circles) in the system.

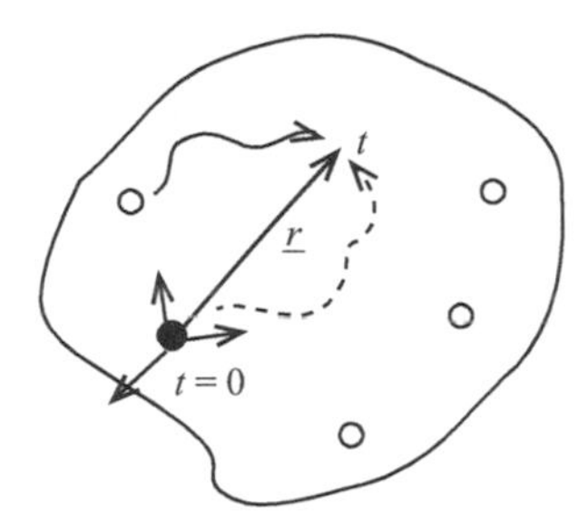

Figure 1.1
Schematic of a system of N particles in which a particle has moved from its initial position at the center of the coordinate system to a position subtended by the vector $\underline{r}$ during a time interval t. Figure created by Sidney Yip.

The initial values of time correlation functions also have simple meanings and relations to structural property of the fluid. Setting $t = 0$ in the definitions of G and G_s, we find

$$G(\underline{r}, 0) = \delta(\underline{r}) + n[g(r) - 1] \tag{1.8}$$

where the first term is the initial value of $G_s(\underline{r}, 0)$, and $g(r)$ is the radial distribution function in equilibrium statistical theory of fluids.

We note just as $g(r)$ can be measured by neutron and x-ray diffraction, G and G_s can be directly measured by neutron inelastic scattering and Rayleigh light scattering (Boon and Yip 1980). Neutron inelastic scattering experiments actually give the double Fourier transforms of the two density correlation functions,

$$S(k, \omega) = \int_{-\infty}^{\infty} dt \int d^3r e^{i(\underline{k} \cdot \underline{r} - \omega t)} G(\underline{r}, t) \tag{1.9}$$

and the same relation exists between $S_s(k, \omega)$ and $G_s(\underline{r}, t)$. $S(k, \omega)$ is generally called the dynamic structure factor, aptly named since the Fourier transform of $g(r)$,

$$S(\underline{k}) = 1 + n \int d^3r e^{i\underline{k} \cdot \underline{r}} [g(r) - 1] \tag{1.10}$$

is called the static structure factor in diffraction studies of fluids.

To see the connection between equations (1.9) and (1.10), we introduce another quantity by rewriting equation (1.9) as

$$S(k, \omega) = \int_{-\infty}^{\infty} dt e^{-i\omega t} F(k, t) \tag{1.11}$$

The function F so defined is known as the intermediate scattering function. It is just the Fourier spatial transform of $G(r, t)$. For isotropic systems such as a simple fluid, G depends only on the magnitude of $\underline{r}$, then

$$F(k,t) = \int d^3 r e^{i\underline{k}\cdot\underline{r}} G(r,t)$$
$$= < n_k^*(0) n_k(t) > -(2\pi)^3 \delta(\underline{k}) \quad (1.12)$$

with

$$n_k(t) = \frac{1}{\sqrt{N}} \sum_{i=1}^{N} e^{i\underline{k}\cdot\underline{R}_i(t)} \quad (1.13)$$

We can ask what is the initial value of $F(k, t)$. At $t = 0$, equation (1.12) gives

$$F(k,0) = \int d^3 r e^{i\underline{k}\cdot\underline{r}} G(r,0) \quad (1.14)$$

where

$$nG(r,0) = < \sum_{\ell,\ell'} \delta(\underline{r}' - \underline{R}_\ell(0)) \delta(\underline{r} - \underline{R}_{\ell'}(0)) > -n^2 \quad (1.15)$$

Setting $\underline{r}' = 0$ without any loss of generality, we can write out the double sum explicitly

$$nG(r,0) = N < \delta(\underline{R}_1)\delta(\underline{r} - \underline{R}_1) > + N(N-1) < \delta(\underline{R}_1)\delta(\underline{r} - \underline{R}_2) > -n^2$$
$$= \frac{N}{Q_N} \int d^3 R_1 \ldots d^3 R_N e^{-\beta U} \delta(\underline{R}_1) \delta(\underline{r})$$
$$+ \frac{N(N-1)}{Q_N} \int d^3 R_1 \ldots d^3 R_N e^{-\beta U} \delta(\underline{R}_1) \delta(\underline{r} - \underline{R}_2) - n^2$$
$$= \frac{N}{V} \delta(\underline{r}) + n^2 g(r) - n^2 \quad (1.16)$$

Inserting this result into equation (1.14) gives

$$F(k,0) = 1 + n \int d^3 r e^{i\underline{k}\cdot\underline{r}} [g(r) - 1] \equiv S(k) \quad (1.17)$$

Hence the basic information in the density correlation functions is the time-dependent particle positions.

There are different theoretical levels at which one can analyze $G(r, t)$. Each is based on a certain level of theoretical rigor. Equivalently, each level is appropriate for a corresponding range of space-time variations (or frequencies and wavelengths) of the thermal fluctuations of interest (Yip and Boon 2003).

Hydrodynamics

This method of analyzing time correlation functions is based on the solution of the linearized equations of hydrodynamics (continuum fluid dynamics or Navier–Stokes equations) as an initial-value problem. For G_s, the corresponding hydrodynamic description is the time-dependent diffusion equation. Results from the hydrodynamic equations are valid at long wavelengths and low frequencies because these are the conditions under which the continuum description is appropriate (Boon and Yip 1980).

Consider the system of linearized equations of hydrodynamics describing the fluctuations (deviation from equilibrium) in number density, velocity, and temperature

$$\frac{\partial}{\partial t}\delta\rho(\underline{r},t)+\rho_o\underline{\nabla}\cdot\underline{v}(\underline{r},t)=0 \tag{1.18}$$

$$\left[\frac{\partial}{\partial t}-\nu_1(\nabla^2+\underline{\nabla}\underline{\nabla})-\nu_2\underline{\nabla}\underline{\nabla}\right]\underline{v}(\underline{r},t)+\gamma^{-1}c^2\left[\frac{1}{\rho_o}\underline{\nabla}\delta\rho(\underline{r},t)+\alpha\underline{\nabla}\delta T(\underline{r},t)=0\right] \tag{1.19}$$

$$\left[\frac{\partial}{\partial t}-\frac{1}{\rho_o C_V}\kappa\nabla^2\right]\delta T(\underline{r},t)+\frac{1}{\alpha}(\gamma-1)\underline{\nabla}\cdot\underline{v}(\underline{r},t)=0 \tag{1.20}$$

where $\delta\rho=\rho-\rho_o$, $\delta\underline{v}=\underline{v}$, and $\delta T=T-T_o$, ν_1, ν_2 are the kinematic viscosities, $\gamma=C_P/C_V$ is the specific heat ratio, c the sound speed, α the thermal expansivity, and κ the thermal conductivity. One can solve these equations for the double Fourier transform of $\delta\rho(\underline{r},t)$, subject to the initial conditions, $\delta\rho(\underline{r},0)=\delta(\underline{r})$, $\delta\underline{v}(\underline{r},0)=0$, and $\delta T(\underline{r},0)=0$. The result for $S(k,\omega)$ takes the form

$$S(k,\omega)=n\left(\frac{\partial n}{\partial p}\right)_T\left[(1-C_V/C_P)\frac{D_2k^2}{\omega^2+(D_2k^2)^2}+\frac{C_V}{C_P}\frac{D_1k^2ck^2}{(\omega^2-c^2k^2)^2+(D_1\omega k^2)^2}\right]$$
$$-n\left(\frac{\partial n}{\partial p}\right)_T(1-C_V/C_P)\frac{D_2k^2(\omega^2-c^2k^2)}{(\omega^2-c^2k^2)^2+(D_1\omega k^2)^2} \tag{1.21}$$

where $D_1=(\eta+4\eta_s/3)/mn$ and $D_2=\kappa/mnC_P$, with η, η_s being the bulk and shear viscosity coefficients.

We see that the dynamic structure factor given by the equations of hydrodynamics consists of three terms. The first term is a Lorentzian line shape centered at the origin (zero frequency), the second term is a displaced Lorentzian line shape centered at the sound-wave frequency $\omega=ck$, with c being the adiabatic sound speed, and the third term has a negative sign and seems to be a combination of the first two. Indeed, physically, the first term represents the diffusion of temperature fluctuations, while the second term

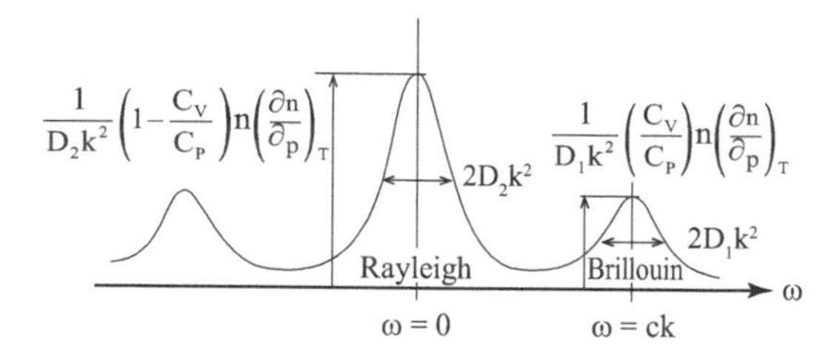

Figure 1.2
Schematic of the dynamic structure factor of a liquid at long wavelengths where the frequency spectrum can be accurately described by linearized hydrodynamics. This line shape is well known as the hydrodynamic limit of the density correlation function. Figure created by Sidney Yip.

represents pressure (sound) propagation. These are the two "normal modes" of thermal fluctuations in the fluid. This is reasonable since a density fluctuation should result in fluctuations in temperature and pressure just through the equation of state. The third term is an interference term between the temperature and pressure fluctuations; it can be destructive or constructive, depending on the difference between ω and ck in the numerator (constructive at low frequencies and destructive at high frequencies).

Figure 1.2 is a sketch of the three-line spectrum of density fluctuations described by equation (1.21). The central peak (fluctuations in temperature) is known as the Rayleigh component; its width is proportional to the thermal conductivity. The two symmetric side peaks are called the Brillouin doublet. Their positions give the sound speed and their width is proportional to the sound attenuation coefficient D_1. This kind of spectral features is well known in light scattering measurements on simple liquids.

The hydrodynamic description of fluid fluctuations is expected to break down at finite values of (k, ω). Qualitatively this condition can be expressed as $kl > 1$, or $kl < 1$, and $\omega > \omega_c$ or $\omega < \omega_c$, where l and ω_c are the collision mean free path and collision frequency, respectively. We can extend the hydrodynamic description by allowing the thermodynamic parameters in equations (1.19) and (1.20) to become wavenumber dependent and the transport coefficients to be wavenumber and frequency dependent. This approach is conventionally known as generalized hydrodynamics (Yip and Boon 2003).

The basic idea of generalized hydrodynamics can be illustrated by working out explicitly the case of transverse current correlation. One of the fundamental differences between simple liquids and solids is that the former cannot support a shear stress—a liquid would simply flow under shear. Another way of saying it is the shear modulus of a liquid should be zero (see essay 4 in connection with the stability criterion for melting of a solid). On the other hand, it is also known that at sufficiently short wavelengths or high frequencies, shear waves can propagate through a simple liquid because then the system behaves like a viscoelastic medium. According to hydrodynamics, we know the frequency spectrum of the transverse current correlation describes a diffusive process at all frequencies. Therefore, the absence of a propagating mode is due to the inability of linearized hydrodynamics to treat viscoelastic behavior at finite k and ω.

The hydrodynamic description of fluctuations in fluids is expected to break down at finite values of (k, ω) when $kl \gtrsim 1$ and $\omega \gtrsim \omega_c$, where l is the collision mean free path and ω_c the mean collision frequency. Nevertheless, we can extend the hydrodynamic description by allowing the thermodynamic coefficients in equations (1.19) and (1.20) to become wavenumber dependent and the transport coefficients to become k- and ω-dependent. To give an example of the generalized hydrodynamics approach, we extend equation (1.20) by writing

$$\frac{\partial}{\partial t} J_t(k,t) = -k^2 \int_0^t dt' K_t(k, t-t') J_t(k,t') \tag{1.25}$$

The kernel $K_t(k, t)$ is called a memory function; it is itself a time correlation function like $J_t(k, t)$. The role of K_t is to enable J_t to take on a short-time behavior that is distinctly different from its behavior at long times. It is reasonable that a quantity such as $K_t(k, t)$ should be present in the extension of hydrodynamics. With the introduction of a suitable $K_t(k, t)$, we expect equation (1.25) will be able to describe shear wave propagation at finite (k, ω), a signature of viscoelasticity. Thus, we can impose the conditions

$$K_t(k, t=0) = (nm)^{-1} G_\infty(k) \tag{1.26}$$

and

$$\lim_{k \to 0} \int_0^\infty dt\, K_t(k,t) = \nu \tag{1.27}$$

where $G_\infty(k)$ and ν are the high-frequency shear modulus and kinematic viscosity respectively. Both G_∞ and ν are properties of the liquid,

$$\frac{(kv_0)^2}{nm} G_\infty(k) = \frac{1}{2\pi} \int_{-\infty}^{\infty} d\omega\, \omega^2 J_t(k,\omega) \tag{1.28}$$

$$2v_0^2 \nu = \lim_{\omega \to 0} \lim_{k \to 0} \left(\frac{\omega}{k}\right)^2 J_t(k,\omega) \tag{1.29}$$

Moreover, equation (1.28) can be reduced to a kinetic contribution $(kv_0^2)^2$ and an integral over $g(r)$ and potential function derivative that can be evaluated by quadrature.

Equations (1.26) and (1.27) may be regarded as constraints or "boundary conditions" on $K_t(k, t)$. By themselves they are not sufficient to determine the memory function.

Empirical forms have been proposed for $K_t(k, t)$ with adjustable parameters determined by imposing equations (1.26) and (1.27). A common practice is to postulate a single relaxation time model

$$K_t(k, t) = [G_\infty(k)/nm] \exp[-t/\tau(k)] \tag{1.30}$$

where we are still free to specify the wavenumber-dependent relaxation time $\tau(k)$. Notice that equation (1.26) has already been incorporated. Applying equation (1.27) we obtain $\tau(k = 0) = nmv/G_\infty(0)$, a quantity sometimes called the Maxwell relaxation time in viscoelastic theories. Furthermore, we expect $\tau(k)$ to be a decreasing function of k on the grounds that fluctuations at shorter wavelengths generally dissipate more rapidly. The simple interpolation expression

$$\frac{1}{\tau_t^2(k)} = \frac{1}{\tau_t^2(0)} + (kv_0)^2 \tag{1.31}$$

would be consistent with this expectation and entails no further parameters. There exist more elaborate models for $\tau(k)$ as well as for $K_t(k, t)$, but the model equation (1.30) with equation (1.31) has the virtue of simplicity. Then equation (1.25) gives

$$\begin{aligned} J_t(k, \omega) = {} & \frac{2v_0^2 k^2 K_t(k, 0)}{\tau_t(k)} \\ & \times \left\{ \left[\omega^2 - \left\{ k^2 K_t(k, 0) - \frac{1}{2\tau_t^2(k)} \right\} \right]^2 \right. \\ & \left. + \left[k^2 K_t(k, 0) - \frac{1}{4\tau_t^2(k)} \right] \Big/ \tau_t^2(k) \right\}^{-1} \end{aligned} \tag{1.32}$$

The effect of the memory function is seen in the spectral behavior of $J_t(k, \omega)$. Whenever

$$k^2 K_t(k, 0) > \frac{1}{2\tau_t^2(k)} \tag{1.33}$$

there will exist a finite frequency, where the denominator in equation (1.32) is a minimum, and $J_t(k, \omega)$ will show a resonant peak. The resonant structure indicates a propagating mode associated with shear waves. Notice that equation (1.33) cannot hold at sufficiently small k; thus, in the long wavelength limit, equation (1.32) can only describe diffusion.

Generalized hydrodynamic descriptions for other time correlation functions have also been proposed using memory function equations such as equation (1.25). We will summarize here the results for density and longitudinal current fluctuations. The continuity equation, equation (1.6), is an exact expression, unlike the Navier–Stokes or the energy transport equation. One of its implications is a rigorous relation between the density correlation function, $F(k, t)$, and the longitudinal current correlation function $J_t(k, t)$. In terms of the dynamic structure factor $S(k, \omega)$, the relation is

$$J_l(k, \omega) = (\omega/k)^2\, S(k, \omega) \tag{1.34}$$

Since this holds in general, we will focus our attention on $J_l(k, \omega)$. For purposes of illustration, we assume temperature fluctuations can be ignored. This means we are assuming the isothermal approximation and equation (1.34) becomes

$$J_l(k,\omega) = 2v_0^2 \frac{(\omega k)^2}{[\omega^2 - (c_T k)^2]^2 + [\omega \nu k^2]^2} \tag{1.35}$$

with $c_T = c_0/\nu$ being the isothermal sound speed. We see, in the hydrodynamic description, the longitudinal current fluctuations, in contrast to the transverse current fluctuations, propagate at a frequency essentially given by to $\omega \sim c_T k^2$. If temperature fluctuations were not neglected, the propagation frequency would be $c_0 k$ and the damping constant governed by the sound attenuation coefficient Γ as in equation (1.21) instead of ν as in equation (1.35).

The inadequacy of the hydrodynamic description equation (1.35) at finite (k, ω) values is more subtle than is the case of $J_t(k, \omega)$. We find equation (1.35) gives an overestimate of the damping of fluctuations, and it does not describe any of the effects associated with the intermolecular structure as manifested through the static structure factor $S(k)$. The extension of equation (1.35) can proceed by writing

$$\frac{\partial J_l(k,t)}{\partial t} = -\int_0^t dt' K_l(k, t-t') J_l(k,t') \tag{1.36}$$

with

$$K_l(k,t) = \frac{(kv_0)^2}{S(k)} + k^2 \phi_l(k,t) \tag{1.37}$$

The form of $K_l(k, t)$ is motivated by the coupling of equations (1.10) and (1.11), and the generalization of the isothermal compressibility x_T, $nk_B T x_T \to S(k)$. Combining equations (1.35) and (1.36) gives

$$J_l(k,\omega) = \frac{2v_0^2 (\omega k)^2 \phi_l'(k,\omega)}{\left[\omega^2 - \frac{(kv_0)^2}{S(k)} + \omega k^2 \phi_l''(k,\omega)\right]^2 + \left[\omega k^2 \phi_l'(k,\omega)\right]^2} \tag{1.38}$$

where ϕ_l'' and ϕ_l' are the real and imaginary parts of

$$\phi_l(k,s) = \int_0^\infty dt\, e^{-st}\phi_l(k,t) \tag{1.39}$$

with $s = i\omega$. They describe the dissipative and reactive responses, respectively.

It is evident from a comparison of equation (1.38) with equation (1.35) that in addition to the generalization of the isothermal compressibility, the longitudinal viscosity has become a complex k- and ω-dependent quantity. Through $\phi_l(k, t)$ we can again introduce physical models and use various properties to determine the k dependence. One way to characterize the breakdown of hydrodynamics in the case of $J_l(k, \omega)$ is to follow the frequency of the propagating mode as k increases. Notice first that by virtue of equation (1.34), $J_l(k, \omega)$ always shows a peak at a nonzero frequency. At small k, this peak is associated with sound propagation. If we define

$$c(k) = \frac{\omega_m(k)}{k} \tag{1.40}$$

where $\omega_m(k)$ is the peak position, then $c(k)$ in the long wavelength limit is the adiabatic sound speed. This being the case, it is reasonable to regard equation (1.40) as the speed at which collective modes propagate in the fluid at any wavenumber. In terms of $c(k)$, we have a well-defined quantity for discussing the variation of propagation speed at finite k. Notice that we do not refer to the propagating fluctuations at finite k as sound waves, because the latter are excitations that manifest clearly in $S(k, \omega)$.

There exist computer simulation results and neutron inelastic scattering data on simple liquids from which $c(k)$ can be determined. Figure 1.3 shows a comparison of these results with a generalized hydrodynamics calculation. Also shown are the adiabatic sound speed $c_0(k)$ and the high-frequency sound speed $c_\infty(k)$

$$c_0(k) = v_0[\gamma/S(k)]^{1/2} \tag{1.41}$$

$$c_\infty(k) = \left\{\frac{1}{nm}\left[\frac{4}{3}G_\infty(k) + K_\infty(k)\right]\right\}^{1/2} \tag{1.42}$$

where K_∞ is the high-frequency bulk modulus. It is seen in equation (1.40) that $c_0(k)$ and $c_\infty(k)$ provide lower and upper bounds on $c(k)$. The fact that $c(k)$ deviates from both may be attributed to dynamical effects, which cannot be described through static properties such as in equations (1.41) and (1.42). Relative to the adiabatic sound speed $c_0(k \to 0)$, we see in $c(k)$ first an enhancement as k increases up to about 1 A^{-1} then a sharp decrease at larger k. The former behavior, a positive dispersion, is due to shear relaxation, whereas the latter, a strong negative dispersion, is due to structural correlation

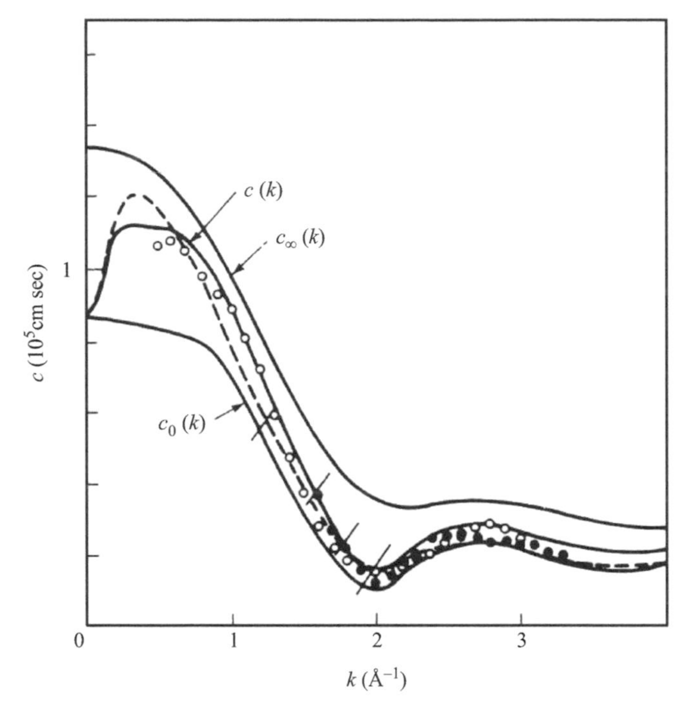

Figure 1.3
Variation of propagation velocities with wavenumber in liquid argon. Generalized hydrodynamics results are given as the solid curve denoted by $c(k)$ and by the dashed curve, neutron scattering measurements are denoted by the closed circles and slash marks, and computer simulation data are denoted by open circles. The quantities $c_0(k)$ and $c_\infty(k)$ are defined in the text (Yip 1976). Figure from (Yip 1976).

effects represented by $S(k)$. From this discussion, we may conclude that an expression such as equation (1.38), with rather simple physical models for $\phi_l(k, t)$, provides a semi-quantitatively correct description of density and current fluctuations at finite (k, ω).

Kinetic Theory

To extend the continuum description to the molecular level, we introduce the distribution function $f(\underline{r}, \underline{v}, t)$ whose velocity integral gives the number density $\rho(\underline{r}, t)$. For gases and liquids, one can derive transport or kinetic equations governing the evolution of the distribution function, an example being the linearized Boltzmann equation in the kinetic theory of gases. It can be shown the density correlation can be defined as (van Leeuwen and Yip 1965)

$$G(r,t) = \int d^3v f(\underline{r}, \underline{v}, t) \tag{1.43}$$

where f is the solution to a kinetic equation subject to the initial condition

$$f(\underline{r}, \underline{v}, 0) = \delta(\underline{r}) M(\underline{v}) \tag{1.44}$$

where M is the Maxwellian distribution of particle velocity.

In the theory of particle and radiation transport in fluids, there exists a well-established connection between the continuum approach as represented by the hydrodynamics equations and the molecular approach as represented by kinetic equations in phase space, an example of which is the Boltzmann equation in gas dynamics. Through this connection, we can obtain expressions for calculating the input parameters in the continuum equations, such as the transport coefficients. We can also solve the kinetic equations directly to analyze thermal fluctuations at finite (k, ω), and in this way take into account, explicitly, the effects of spatial correlations and detailed dynamics of molecular collisions. In contrast to generalized hydrodynamics, the kinetic theory method allows us to derive, rather than postulate, the space-time memory functions like $K(k, t)$.

The essence of the kinetic theory description is that particle motions are followed in both configuration and momentum space. Analogous to the previous definition of space-time density correlation, we begin with the phase-space density (Yip 1979)

$$A(\mathbf{rp}t) = \sum_{i=1}^{N} \delta(\mathbf{r} - \mathbf{R}_i(t)) \delta(\mathbf{p} - \mathbf{P}_i(t)) \tag{1.45}$$

and the time-dependent phase-space density correlation function in equation (1.7)

$$C(\mathbf{r} - \mathbf{r}', \mathbf{pp}', t) = \langle \delta A(\mathbf{rp}t) \delta A(\mathbf{r}'\mathbf{p}'0) \rangle \tag{1.46}$$

with $\langle A \rangle = n f_0(p)$. The fundamental quantity in the analysis is now $C(\mathrm{r}, \mathrm{pp}', t)$, from which all previous space-time correlation functions can be obtained by appropriate integration over the momentum variables. For example

$$G(\mathbf{r}, t) = \int d^3p d^3p' C(\mathbf{r}, \mathbf{pp}', t) \tag{1.47}$$

Various methods have been proposed to derive the equation governing $C(\mathrm{r}, \mathrm{pp}', t)$. All the results can be put into the generic form

$$\left(z - \frac{\mathbf{k} \cdot \mathbf{p}}{m} \right) C(k\mathbf{pp}'z) - \int d^3p'' \phi(k\mathbf{pp}''z) C(k\mathbf{p}''\mathbf{p}'z) = -iC_0(k\mathbf{pp}') \tag{1.48}$$

where

$$C(k\mathbf{pp}'z) = \int d^3r \int_0^\infty dt\, e^{i(\mathbf{k} \cdot \mathbf{r} - zt)} C(\mathbf{r}, \mathbf{pp}', t) \tag{1.49}$$

with the initial condition

$$C_0(k\mathbf{pp}') = \int d^3 r e^{i\mathbf{k}\cdot\mathbf{r}} C(\mathbf{r}, \mathbf{pp}', t = 0)$$
$$= n f_0(p)\delta(\mathbf{p} - \mathbf{p}') + n^2 f_0(p) f_0(p') h(k) \tag{1.50}$$

and $nh(k) = S(k) - 1$. In equation (1.48), the function $\phi(k\mathrm{pp}'z)$ is the phase-space memory function, which plays the same role as the memory function $K(k, t)$ in equation (1.25) or equation (1.36). It contains all the effects of molecular interactions. If ϕ were identically zero, then equation (1.48) would describe a noninteracting system in which the particles move in straight line trajectories at constant velocities. We can also think of ϕ as the collision kernel in a transport equation.

There are a number of formal properties of ϕ pertaining to symmetries, conservation laws, and asymptotic behavior, which one can analyze. Also, explicit calculations have been made under different conditions, such as low density, weak coupling, or relaxation time models. In general, it is useful to separate ϕ into an instantaneous (or static) part and a time-varying (or collisional) part

$$\phi(k\mathbf{pp}'z) = \phi^{(s)}(k\mathbf{p}) + \phi^{(c)}(k\mathbf{pp}'z) \tag{1.51}$$

where

$$\phi^{(s)}(k\mathbf{p}) = -\frac{\mathbf{k}\cdot\mathbf{p}}{m} n f_0(p) C(k) \tag{1.52}$$

The quantity $C(k) = (S(k) - 1)/nS(k)$ is known as the direct correlation function. Physically, $\phi^{(s)}$ represents the effects of mean-field interactions with $nC(k)$ as the effective potential of the fluid system.

The calculation of $\phi^{(c)}$ is a difficult problem because we have to deal explicitly with the details of collision dynamics. It can be shown that in the limit of low densities, low frequencies, and small wavenumbers, $\phi^{(c)}$ reduces to the collision kernel in the linearized Boltzmann equation. This connection is significant because the Boltzmann equation is the fundamental equation in the study of transport coefficients and of the response of a gas to external perturbations.

The basic assumption underlying the Boltzmann equation is that intermolecular interactions can be treated as a sequence of uncorrelated binary collisions. This assumption renders the equation much more tractable, but it also limits the validity of the equation to low-density gases. Figure 1.4 shows the frequency spectrum of density fluctuations in xenon gas at 349.6 K and 1.03 atmosphere calculated according to the procedure

$$S(k, \omega) = \frac{1}{\pi} Re \int d^3 p d^3 p' C(k\mathbf{pp}'z)_{z = i\omega} \tag{1.53}$$

where C is determined from equation (1.51) with $\phi^{(c)}$ given by the binary collision kernel for hard sphere interactions. At such a low density, it is valid to ignore $\phi^{(s)}$ and the second term in equation (1.50).

Also shown in figure 1.4 are the experimental data from light scattering spectroscopy. The good agreement is evidence that the linearized Boltzmann equation provides an accurate description of thermal fluctuations in low-density gases in the kinetic regime where $kl \sim 1$. The agreement is less satisfactory when the data are compared with the results of hydrodynamics; in this case the calculated spectrum shows essentially no structure. This again indicates that at finite (k, ω), hydrodynamics theory overestimates the damping of density fluctuations. Generally speaking, kinetic theory calculations have been quantitatively useful in the analysis of light scattering experiments on gases and gas mixtures.

For moderately dense systems, typically fluids at around the critical density, the Boltzmann equation needs to be modified to take into account the local structure of the fluid. In the case of hard spheres, the modified equation that is adopted is the Enskog equation, which involves $g(\sigma)$, the pair distribution function at contact (with σ the hard sphere diameter); the collision term differs from the collision integral in the linearized

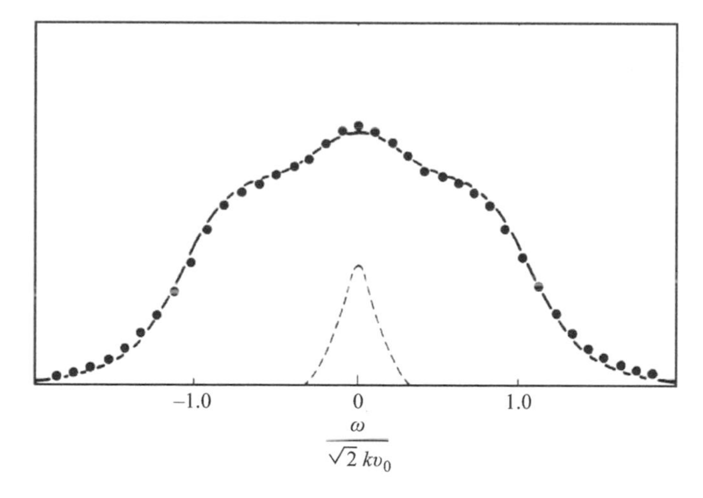

Figure 1.4
Frequency spectrum of dynamic structure factor in xenon gas at 349.6 K and 1.03 atmosphere; light scattering data for 6,328 Å incident light; and scattering angle of 169.4° are shown as closed circles, while the full curve denotes results obtained using the linearized Boltzmann equation for hard spheres. Calculated spectrum has been convolved with the resolution function shown by the dashed curve (Sugawara, Yip, and Sirovich 1968). Reprinted figure with permission from Sugawara, Yip, and Sirovich (1968). Copyright 1968 by the American Physical Society.

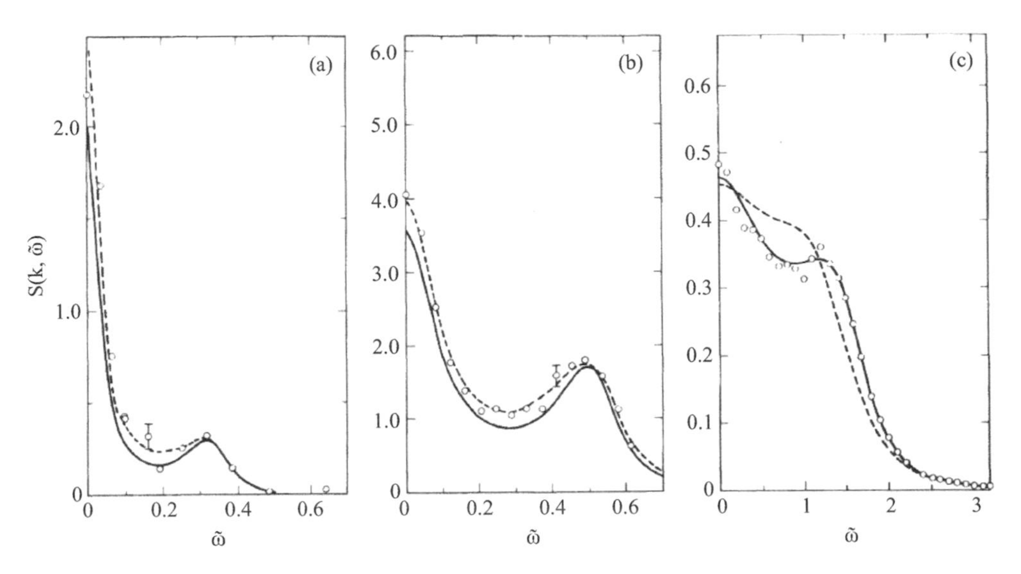

Figure 1.5
Dynamic structure factor of hard-sphere fluids at low wavenumbers and three densities, (a) V/Vo = 1.6, $k\sigma = 0.38$, (b) V/Vo = 3, $k\sigma = 0.44$, and (c) V/Vo = 10, $k\sigma = 0.41$. MD results (circles) are compared with kinetic theory (solid curves) and hydrodynamic (dashed curves) calculations (Alley 1983). Reprinted figure with permission from (Alley, Alder, and Yip 1983). Copyright 1983 by the American Physical Society.

Boltzmann equation for hard spheres only in the presence of two-phase factors, which represent the nonlocal spatial effects in collisions between molecules of finite size.

Figure 1.5 shows the frequency spectra of density fluctuations obtained from simulation and kinetic theory at rather long wavelengths in hard sphere fluids and at three densities, corresponding roughly to liquid density at the triple point, around the critical density, and dilute gas density (Alley, Alder, and Yip 1983). The $k\sigma$ values are such that using the expression $l^{-1} = \sqrt{2}\pi n\sigma^2 g(\sigma)$ for the collision mean free path, we find that for the three cases (a)–(c) a molecule on the average would have suffered about twenty, five, and one collisions, respectively, in traversing a distance equal to the hard sphere diameter. On this basis, we might expect the spectra in (b) and (c) to be dominated by hydrodynamic behavior, while the spectra in (a) should show significant deviations.

The kinetic theory curves in figure 1.5 are kinetic model solutions to the generalized Enskog equation. They are seen to quantitatively describe the computer simulation data. We could have expected good agreement in the lowest density case, which is nevertheless two orders of magnitude higher in density than a gas under standard conditions. That the theory is still accurate at condition (b) is already somewhat unexpected.

So it seems rather surprising that a kinetic theory that treats the interactions as only uncorrelated binary collisions is applicable at liquid density, as shown in (a).

Although not immediately recognized at the time, a plausible explanation is that the dynamic structure factor is a property that reflects the dynamics of particle collisions only in an average sense. Specific details of correlated versus uncorrelated binary collisions are not coupled strongly with fluctuations in the density. This is why a theory without correlated binary collisions is sufficient to account for the spectral features seen in figure 1.6. On the other hand, there are *other* correlation functions one can examine where the effects of correlated collisions can make a qualitative difference. As an example, suppose we consider the time-dependent behavior of the transverse current correlation. In this case, we expect the backscattering effects from correlated collisions to give rise to a temporal relaxation of the correlation that goes *negative*. Then we would expect kinetic theory without correlated binary collisions would not be able to capture this behavior. Indeed, this can be deduced from figure 1.6.

The failure of the generalized Enskog equation to account properly for the simulation results at low frequencies can be traced to the presence of a slower decaying component in the data for $F(k, t)$. It seems reasonable to associate this with the relaxation of clusters of particles, which should become important at high densities. Just like the onset of shear wave propagation, this characteristic feature is part of the viscoelastic behavior expected of dense fluids. In order to describe such effects in the present context, it is now recognized that correlated collisions will have to be included in the kinetic equation. Aside from density and thermal fluctuations, it is also known that the transport coefficients derived from the Enskog equation are in error up to a factor of 2 at the liquid density when compared to computer simulation data on hard spheres. Moreover, simulation studies have revealed a nonexponential, long-time decay of the velocity autocorrelation function that cannot be explained by the Enskog theory.

Any attempt to treat correlated collision effects necessarily leads to nonlinear kinetic equations. For practical calculations, it appears that only the correlated binary collisions, called ring collisions, are tractable. To incorporate these dynamical processes in the kinetic theory, we can develop a formalism wherein $\phi^{(c)}$ is given as the sum $\phi^{(c)} = \phi_E + \phi_R$, where ϕ_E is the memory function for the generalized Enskog equation, and ϕ_R describes the ring collision contribution. In essence ϕ_R can be expressed schematically as $\phi_R = VCCV$, where V is an effective interaction, which involves the actual intermolecular potential and the equilibrium distribution function of the fluid, and C is the phase space correlation function. The important point to note is that the memory function now depends quadratically on C, thereby making equation (1.50) a nonlinear kinetic equation. The appearance of nonlinearity, or feedback effects, is not so surprising when we recognize

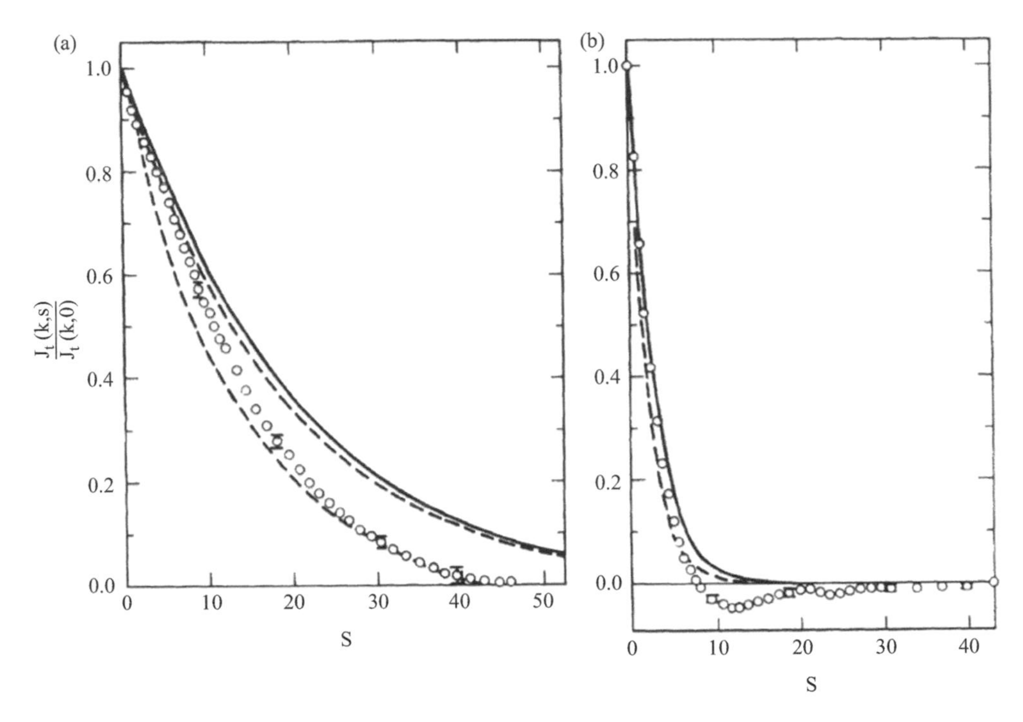

Figure 1.6
Normalized transverse current correlation of a hard-sphere fluid at V/Vo = 1.6 at two wavelengths, $k\sigma = 0.76$ (panel a) and 2.28 (panel b) respectively. MD results (circles) are compared with kinetic theory (solid curves) and hydrodynamics (dashed curves) calculations. In the hydrodynamic results (panel a), the upper dashed curve corresponds to the Enskog approximation for the viscosity while the lower dashed curve corresponds to using the correct viscosity for hard spheres (Alley 1983). Reprinted figure with permission from Alley, Alder, and Yip (1983). Copyright 1983 by the American Physical Society.

that in a dense medium the motions of a particle will have considerable effects on its surroundings, which in turn will react and influence the subsequent motions of the particle. The inclusion of correlated collisions is a significant development in the study of transport phenomena called renormalized kinetic theory.

The presence of ring collisions unavoidably makes the analysis of time correlation functions considerably more difficult. Nevertheless, it can be shown analytically that we obtain a number of nontrivial collective properties characteristic of a dense fluid, such as a power law decay of the velocity autocorrelation function, and nonanalytic density expansions of sound dispersion and transport coefficients.

Another way to extend the range of the hydrodynamic description is to replace the transport coefficient by an appropriate transport function, which is an integral operator. A simple example of this procedure is to take the diffusion equation

$$\frac{d}{dt}F_s(k,t) = -Dk^2 F_s(k,t) \tag{1.54}$$

and write it as

$$\frac{d}{dt}F_s(k,t) = -k^2 \int_0^t K(t-t')F_s(k,t')dt' \tag{1.55}$$

The kernel K is called the memory function because it couples the behavior of dF_s/dt to the behavior of F_s at earlier times. If the medium happens to have no memory effects, then we would have

$$K(t-t') = D\delta(t-t') \tag{1.56}$$

and equation (1.55) reduces back to equation (1.54).

Mesoscale in Liquid Dynamics

The essence of the opening essay in this work is the dual concept of time correlation functions and the use of memory function to describe the complexity of collisional dynamics in fluids. This theme is continued in essay 2 where atomistic simulations emerged to be a powerful capability for probing molecular mechanisms. The memory function approach is further highlighted in essay 3 to treat nonlinear coupling dynamics in the case of particle localization in diffusion. Collectively the three essays in part I point to an early recognition of a research domain where microscale and macroscale behavior both need to be addressed. In liquid-state collisional dynamics, the appropriate formalism is called kinetic theory. But the underlying idea is equally applicable to solids and even glasses. With hindsight, this could be regarded as a natural precursor to the development of the mesoscale science research frontier decades later (Yip and Short 2013).

References

Alley, W. E., B. J. Alder, and S. Yip. 1983. "The neutron scattering function for hard spheres." *Physical Review A* 27: 3174.

Boon, J.-P., and S. Yip. 1980. *Molecular Hydrodynamics*. New York: McGraw-Hill. (Reprinted by Dover in 1991.)

Sugawara, A., S. Yip, and L. Sirovich. 1968. "Kinetic theory analysis of light scattering in gases." *Physical Review* 168: 121.

van Leeuwen, J. M. J., and S. Yip. 1965. "Derivation of kinetic equations for slow-neutron scattering." *Physical Review* 139 (4A): 1138.

Yip, S. 1976. "High-frequency short-wavelength fluctuations in fluids." In *New Directions in Physical Acoustics*, edited by D. Sette, 55–96. Amsterdam: North-Holland Publishing Co.

Yip, S. 1979. "Renormalized kinetic theory of dense fluids." *Annual Review of Physical Chemistry* 30: 547.

Yip, S., and J.-P. Boon. 2003. "Molecular hydrodynamics." In *Encyclopedia of Physical Science and Technology*, 3rd ed., 141–159. San Diego: Academic Press.

Yip, S., and M. Short. 2013. "Multiscale materials modelling at the mesoscale." *Nature Materials* 12: 774–777.

Further Reading

This essay is primarily based on the monograph *Molecular Hydrodynamics* (Boon and Yip 1980). The following studies of time correlation functions are also relevant:

Martin, P. C., and S. Yip. 1968. "Frequency-dependent friction constant analysis in simple liquids." *Physical Review* 170: 155. An early model of a memory function.

Chung, C. H., and S. Yip. 1969. "Generalized hydrodynamics and time correlation functions." *Physical Review* 183: 323. Extension of hydrodynamics through space-time correlations.

van Leeuwen, J. M. J., and S. Yip. 1969. "Derivation of kinetic equations for slow-neutron scattering." *Physical Review* 113: A1138. Detailed derivation of the Boltzmann transport equation.

Yip, S. 1971. "Rayleigh scattering in dilute gases." *The Journal of the Acoustical Society of America* 49 (3C): 941. Correlation function calculations for hard-spheres.

Mazenko, G. F., T. Wei, and S. Yip. 1972. "Thermal fluctuations in a hard-sphere gas." *Physical Review* A (6): 1981. Correlation function calculations for hard-spheres.

Boley, C. D., and S. Yip. 1972. "Spectral distribution of light scattered in dilute gases and gas mixtures." *Journal de Physique* 33 (C1). Light scattering in binary gas mixtures.

Yip, S. 1974 "Quasielastic scattering in neutron and laser spectroscopy." In *Spectroscopy in Biology and Chemistry—Neutron, X-Ray, and Laser*, edited by Sow-Hsin Chen and S. Yio, 53. New York: Academic Press. A tutorial on density correlation functions in incoherent neutron scattering and Rayleigh light scattering.

Chen, Sow-Hsin, and S. Yip. 1976. "A new look at neutron molecular spectroscopy." *Physics Today* 29: 32. Theme issue on neutron spectroscopy capabilities for probing molecular systems.

Furtado, P. M., G. F. Mazenko, and S. Yip. 1976. "Kinetic model description of dense hard-sphere fluids." *Physical Review* A (13): 1641. Kinetic model calculations of correlation functions.

Mazenko, G. F., and S. Yip. 1977. "Renormalized kinetic theory of dense fluids." In *Modern Theoretical Chemistry: Statistical Mechanics—Part B*, edited by B. J. Berne, 181. New York: Plenum Press. Kinetic theory with correlated binary collisions.

Leutheusser, E. 1982. "Dynamics of a classical hard-sphere gas. I. Formal theory." *Journal of Physics C: Solid State Physics* 15: 2801. The development of mode-coupling theory for a hard-sphere gas.

Leutheusser, E. 1982. "Dynamics of a classical hard-sphere gas. II. Numerical results." *Journal of Physics C: Solid State Physics* 15: 2827. Numerical results for various time correlation functions.

Yip, S. 2017. "Molecular dynamics of dense fluids: Simulation-Theory Symbiosis." *Proceedings of the Symposium in Honor of Dr. Berni Alder's 90th Birthday* (World Scientific, Singapore) chap. 9, 114.

For classic references on statistical mechanics of gases and liquids, and on time correlation functions readers may wish to consult:

Chapman, S., and T. G. Cowling. 1991. *Mathematical Theory of Non-Uniform Gases*, 3rd ed. Cambridge Univ. Press.

Cercignani, C. 2000. *Rarefied Gas Dynamics*. Cambridge Univ. Press.

Egelstaff, P. A. 1994. *An Introduction to the Liquid State*, 2nd ed. Oxford Univ. Press.

Hansen, J.-P., and I. McDonald. 2013. *Theory of Simple Liquids*, 4th ed. Elsevier.

Martin, P. C. 1970. *Measurements and Correlation Functions*. New York: Gordon and Breach.

Zwanzig, R. 2001. *Nonequilibrium Statistical Mechanics*. Oxford Univ. Press.

2 Atomistic Simulations: A Primer

The motions of atoms or molecules (or particles in a generic sense) in a material assembly can be simulated by numerically solving a set of equations of motion governing all the particles. The method is called molecular dynamics (MD) when the equations are the Newton's equations of motion. These equations are deterministic. A companion method, Monte Carlo (MC), which is stochastic in nature, is also capable of giving the particle positions but not the velocities. Together, they constitute the computational methods appropriate at the atomistic level—see figure P.1 in the prologue. In this essay, we lay out the basic components of MD and emphasize the features that make atomistic simulations unique. Throughout the book, we will encounter MD results that later provide motivations for the method of metadynamics simulation to be introduced in essay 10. Thereafter, in the remaining essays, we will focus on the results of meta-dynamics simulation, which has the capability to reach longer timescales.

Introduction

This essay explains the technique of MD as a particle-tracking method to follow the time evolution of a system of interacting particles, while the companion method, MC, is a statistical method. In MD, the particles move according to the deterministic Newton's equations of motion.

Conceptually, particle tracking is the most intuitive process by which one thinks of transport phenomena in matter. We speak of coefficients of diffusion, viscosity, and thermal conductivity, which, as we have seen in essay 1, are integrals of time correlation functions. For theoretical understanding of liquids, one may begin with the Boltzmann transport equation for dilute gases and its extensions to dense fluids. For condensed matter, such as glasses and solids, the problem becomes more difficult. Extension of the Boltzmann equation to liquid density involves complicated analysis of correlated binary collisions and triple collisions, the mathematical details becoming so heavy that physical insights become obscure. Alternatively, one can turn to particle simulation methods to obtain directly the particle trajectories during the transport process. Then, MD is well suited to the study of dynamical processes, whereas MC can also be used, although in a statistical manner that involves interaction energy rather than intermolecular forces.

Basic Molecular Dynamics

A minimal introduction will need to address the following essential topics: MD definition, pair potential, bookkeeping matters, property calculations, and why MD is unique.

MD is the process of generating the atomic trajectories of a system of N particles by direct numerical integration of Newton's equations of motion, with appropriate specification of an interatomic potential, and appropriate initial and boundary conditions.

Consider a simulation model (the system) of N particles contained in a region of volume V at temperature T, shown schematically in figure 2.1.

The positions of the N particles are specified by a set of N vectors

$$\{\underline{r}(t)\} = (\underline{r}_1(t), \underline{r}_2(t), \ldots, \underline{r}_N(t))$$

where $\underline{r}_j(t)$ denotes the position of particle j at time t. Knowing $\{\underline{r}(t)\}$ at various time instants means we know the particle *trajectories* as they move around. Our model system has energy E, the sum of the kinetic and potential energies of the particles, $E = K + U$

$$K = \frac{1}{2} m \sum_{j=1}^{N} \underline{v}_j^2 \tag{2.1}$$

and U is a prescribed interatomic potential. Notice that U is a function of all the particle positions in the system, $U(\underline{r}_1, \underline{r}_2, \ldots, \underline{r}_N)$, because all particles interact fully with each other. In general, U depends on the positions of all the particles in a complicated fashion. We will soon introduce a simplifying approximation, the assumption of two-body (pairwise) interaction, which makes this important quantity much easier to handle.

To find the atomic trajectories, we solve the Newton's equations of motion

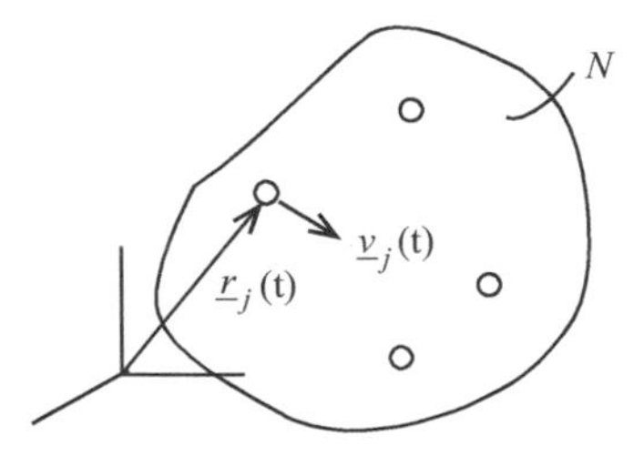

Figure 2.1
Schematic model for MD simulation, a system of N particles at position $\underline{r}_j(t)$ and velocity $\underline{v}_j(t)$, interacting with each other through a potential energy U. Figure created by Sidney Yip.

$$m\frac{d^2\underline{r}_j}{dt^2} = -\underline{\nabla}_{\underline{r}_j}U(\{\underline{r}\}),\ \ j = 1, \ldots, N \tag{2.2}$$

Equation (2.2) is a system of coupled second-order, nonlinear ordinary differential equations. It is to be integrated numerically in MD simulation. We consider a certain time interval divided into many small segments, time step size of Δt. Given the system conditions at some initial time t_o, $\{\underline{r}(t_o)\}$, numerical integration means we advance the system successively by increments of Δt

$$\{\underline{r}(t_o)\} \rightarrow \{\underline{r}(t_o + \Delta t)\} \rightarrow \{\underline{r}(t_o + 2\Delta t)\} \rightarrow \cdots \{\underline{r}(t_o + N_t\Delta t)\} \tag{2.3}$$

where N_t is the number of time steps making up the interval of integration.

To numerically integrate (2.2) for a given U, a simple way is to write a Taylor series expansion

$$\underline{r}_j(t_o + \Delta t) = \underline{r}_j(t_o) + \underline{v}_j(t)\Delta t + \frac{1}{2}\underline{a}_j(t)(\Delta t)^2 + \cdots \tag{2.4}$$

and a similar expansion for $\underline{r}_j(t_o - \Delta t)$. Adding the two expansions gives

$$\underline{r}_j(t_o + \Delta t) = -\underline{r}_j(t_o - \Delta t) + 2\underline{r}_j(t_o) + \underline{a}_j(t_o)(\Delta t)^2 + \cdots \tag{2.5}$$

Notice that the left-hand side is what we want, namely, the position of particle j at the next time step Δt, whereas all the terms on the right-hand side are quantities evaluated at time t_o. We already know the positions at t_o and the time step before, so to use equation (2.5), we need the acceleration of particle j at time t_o. For this, we make use of equation (2.2) and substitute $\underline{F}_j(\{\underline{r}(t_o)\})/m$ in place of the acceleration, where F is just the right-hand side of equation (2.2). Thus, through equation (2.5), one performs the integration in successive time increments, following the system evolution in discrete time steps. Although there are more elaborate ways of doing the integration, this is the basic idea of generating the atomic trajectories, the essence of MD. The particular procedure just described is the Verlet central difference method. Higher accuracy of integration allows one to take a larger value of Δt, which is desirable because one can then cover a longer time interval. On the other hand, the tradeoff is that one needs more memory relative to the simpler method.

The time integrator is at the heart of MD simulation, with the sequence of positions and velocities (trajectories) being the raw output. By far the most computationally intensive part of MD simulation is the force calculation, and in turn the complexity of the force calculation is governed by the prescribed interatomic potential U. In deciding what potential to specify, one strives for the simplest possible description so long as the physics fidelity is not unreasonably compromised. The competition between accuracy of physical description and computational efficacy is inherent to all atomistic calculations.

The pair potential model To make the simulation more tractable, it is common to assume the interatomic potential U can be represented as the sum of two-body interactions

$$U(\underline{r}_1, \ldots, \underline{r}_N) \cong \frac{1}{2} \sum_{i \neq j} V(r_{ij}) \tag{2.6}$$

where r_{ij} is the separation distance between particles i and j. V is the pairwise interaction; it is a central force potential, meaning it is a function only of the separation distance between the two particles, $r_{ij} = |\underline{r}_i - \underline{r}_j|$. A very common two-body interaction energy used in atomistic simulations, known as the Lennard-Jones potential, is

$$V(r) = 4\varepsilon[(\sigma/r)^{12} - (\sigma/r)^6] \tag{2.7}$$

where ε and σ are parameters of the potential. Like all pair potentials, this interaction energy rises sharply (with inverse power of 12) at close interatomic separations, has a minimum, and decays to zero at large separations. See figure 2.2 that also shows the behavior of the interatomic force

$$F(r) = -\frac{dV(r)}{dr} \tag{2.8}$$

as repulsive at short separations and attractive at large separations, arising respectively from the overlap of the electron clouds and the interaction between the induced dipole in each atom. The value of 12 for the first exponent in $V(r)$ has no special significance, as the repulsive term could just as well be replaced by an exponential. The second exponent results from quantum mechanical calculations (the so-called London dispersion force) and therefore is not arbitrary. Regardless of whether one uses equation (2.7) or some other interaction potentials, a short-range repulsion is necessary to give the system a certain size or *volume* (density), without which the particles will collapse onto each other. A longer range attraction is also necessary for the *cohesion* of the system, without which the particles will not stay together as they must in all condensed states of matter. Both are necessary for modeling the physical properties of solids and liquids known from everyday experience.

Bookkeeping matters The simulation system is typically taken to be a cubical cell in which particles are placed either in a regular manner, as in modeling a crystalline lattice, or in some disordered fashion, as in modeling a gas or liquid. The number of particles in the simulation cell was quite small in the early developments because of limited computational resources.

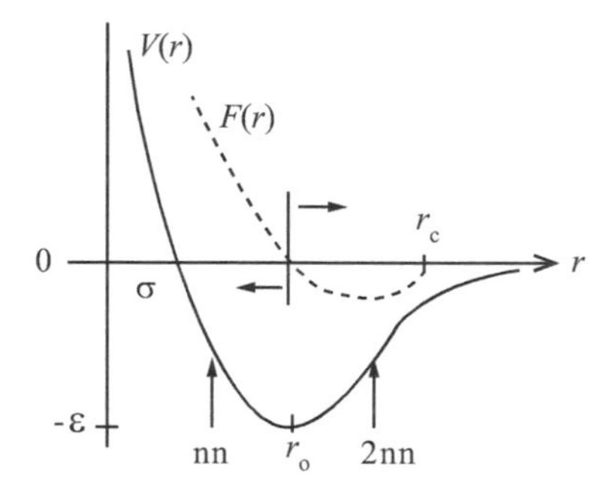

Figure 2.2
Schematic of the Lennard-Jones interatomic potential *V(r)*. The potential vanishes at $r = \sigma$ and has a depth equal to ε. Also shown is the corresponding force *F(r)* between the two particles (dashed curve) which vanishes at r_o. At separations less or greater than r_o the force is repulsive or attractive respectively. Arrows at nn and 2nn indicate typical separation distances of nearest and second nearest neighbors in a solid. Figure created by Sidney Yip.

For simulation of bulk systems (no free surfaces) it is conventional to use the periodic boundary condition (PBC). This means the cubical cell is surrounded by twenty-six identical image cells. For every particle in the simulation cell, there corresponds an image particle in each image cell. The twenty-six image particles move in exactly the same manner as the actual particle, so if an actual particle should happen to move out of the simulation cell, the image particle in the image cell opposite to the exit side will move in. The net effect of imposing PBC is particle number conservation. In other words, N is conserved, and if V is not allowed to change, density n then remains constant.

Since in the pair potential approximation, the particles interact two at a time, a procedure is needed to decide which pair to consider among the pairs between actual particles and between actual and image particles. The *minimum image convention* is a procedure where one treats the nearest neighbor to be an actual particle, regardless of whether this neighbor is an actual particle or an image particle. Another approximation which is useful to keep the computations manageable is to introduce a *force cutoff* distance beyond which particle pairs simply do not see each other (see the force curve in figure 2.2). In order not to have a particle interact with its own image, it is necessary to ensure that the cutoff distance is less than half of the simulation cell dimension.

Another bookkeeping device often used in MD simulation is a neighbor list that keeps track of who are the nearest, second nearest, and so on, neighbors of each particle. This is to save time from checking every particle in the system every time a force calculation is made. The neighbor list can be used for several time steps before updating.

Each update is expensive since it involves $N \times N$ operations for an N-particle system. In low-temperature solids where the particles do not move very much, it is possible to do an entire simulation without or with only a few updating, whereas in simulation of liquids, updating every five or ten steps is quite common.

Property calculations Let $<A>$ denote the time average over the trajectory generated by MD, where A is a dynamical variable, $A(t)$. Two kinds of property calculations are of interest, static properties and time correlation functions. The first is a running time average over the MD trajectories,

$$<A> = \lim_{t \to \infty} \frac{1}{t} \int_{0}^{t} dt' A(t') \tag{2.9}$$

with t taken to be as long as possible. In terms of discrete time steps, equation (2.9) becomes

$$<A> = \frac{1}{N_t} \sum_{k=1}^{N_t} A(t_k) \tag{2.10}$$

where N_t is the number of time steps in the trajectory. The second kind of property is a time-dependent quantity of the form

$$<A(0)B(t)> = \frac{1}{N} \sum_{i=1}^{N} \frac{1}{N_i} \sum_{k=1}^{N_i} A_i(t_k) B_i(t_k + t) \tag{2.11}$$

where B is in general another dynamical variable, and N_i is the number of time origins. Equation (2.11) is called a correlation function of two dynamical variables A and B. Since it is manifestly time-dependent, it is able to represent dynamical information of the system.

Examples of both kinds of averages are commonly encountered in both MD and MC simulations (Allen and Tildesley 1987; Cai et al. 2020)

$$U = < \sum_{i<j}^{N} V(r_{ij}) > \quad \text{potential energy} \tag{2.12}$$

$$T = \frac{1}{3Nk_B} < \sum_{i=1}^{N} m_i \underline{v}_i^2 > \quad \text{temperature} \tag{2.13}$$

$$P = \frac{1}{3V} < \sum_{i=1}^{N} (m_i \underline{v}_i^2 + \underline{r}_i \cdot \underline{f}_i) > \quad \text{pressure} \tag{2.14}$$

$$g(r) = \frac{1}{n 4 \pi r^2 dr} < \sum_{i \neq j}^{N} \delta(r - |\underline{r}_i - \underline{r}_j|) > \quad \text{radial distribution function} \tag{2.15}$$

$$<\Delta r^2> = \frac{1}{N}\sum_{i=1}^{N}[\underline{r}_i(t) - \underline{r}_i(0)]^2 \quad \text{mean-square displacement} \tag{2.16}$$

$$<\underline{v}(0)\cdot\underline{v}(t)> = \frac{1}{N}\sum_{i=1}^{N}\frac{1}{N_i}\sum_{k=1}^{N_i}\underline{v}_i(t_k)\cdot\underline{v}_i(t_k+t) \quad \text{velocity autocorrelation function} \tag{2.17}$$

$$\sigma_{\alpha\beta} = \sum_i\left(\frac{v_a}{V}\right)\sigma^i_{\alpha\beta},\ \sigma^i_{\alpha\beta} = \frac{1}{v_a}<\left\{mv_{i\alpha}v_{i\beta} - \sum_{j>i}\frac{\partial V(r_{ij})}{\partial r_{ij}}\frac{r_{ij\alpha}r_{ij\beta}}{r_{ij}}\right\}> \quad \text{virial stress tensor} \tag{2.18}$$

Writing the stress tensor in the present form suggests the macroscopic tensor can be decomposed into individual atomic contributions, and thus $\sigma^i_{\alpha\beta}$ is known as the atomic level stress at atom *i*. While quite appealing, this is not a general result. The decomposition makes sense only in a homogeneous system where every atom is the same as every other atom.

Properties that make MD unique There is a good deal that should be said about why MD is such a useful simulation technique. Perhaps the most important statement is in this method (consider classical MD for the moment, as opposed to quantum MD) one follows the atomic motions according to the principles of classical mechanics as formulated by Newton and Hamilton. Because of this, the results are physically as meaningful as the potential *U* that is used. One does not have to apologize for any approximation in treating the *N*-body problem. Whatever mechanical, thermodynamic, and statistical mechanical properties that a system of *N* particles should have, they are all still present in the simulation data. Of course how one extracts these properties from the output of the simulation—the atomic trajectories—determines how useful is the simulation. Before any conclusions can be drawn, one should consider how the various properties are to be obtained from the simulation data. We can regard MD simulation as an "atomic video" of the particle motion. There is a great deal of realistic details in the motions themselves, it is up to the viewer to extract the information in a scientifically meaningful way. It is then to be expected that an experienced viewer can extract much more useful information than an inexperienced one!

We enumerate here several basic reasons why MD simulation is useful (or unique). These are meant to guide the thinking of the reader and stimulate discovery and appreciation of the many interesting and thought-provoking aspects of this technique on your own.

1. *Unified study of all physical properties.* By using MD, one can obtain thermodynamic, structural, mechanical, dynamic, and transport properties of a system of particles that can be a solid, liquid, or gas. One can even study chemical and biological properties and reactions that are more difficult and will require using quantum MD.

2. *Several hundred particles are sufficient to simulate bulk matter.* While this is not always true, it is rather surprising that one can get reasonably accurate thermodynamic properties such as equation of state in this way. This is an example that the law of large numbers takes over quickly when one can average over several hundred degrees of freedom.
3. *Direct link between potential model and physical properties.* This is really useful from the standpoint of fundamental understanding of physical matter. It is also very relevant to the structure-property correlation paradigm in materials science.
4. *Complete control over input, initial, and boundary conditions.* This is what gives physical insight into complex system behavior. This is also what makes simulation so useful when combined with experiment and theory.
5. *Detailed atomic trajectories.* This is what one can get from MD, or other atomistic simulation techniques, that experiment often cannot provide. This point alone makes it compelling for the experimentalist to have access to simulation.

We should not leave this discussion without reminding ourselves that there are also significant limitations in MD. The two most important ones are:

Need for sufficiently realistic interatomic potential functions U. This is a matter of what we really know fundamentally about the chemical binding of the system we want to study. Progress is being made in quantum and solid-state chemistry, and condensed-matter physics; these advances will make MD more and more useful in understanding and predicting the properties and behavior of physical systems.

Computational-capability constraints. No computers will ever be big enough and fast enough. On the other hand, things will keep on improving as far as we can tell. Current limits on how big and how long are a billion atoms and about a microsecond in brute force simulation.

Perspective

There are many ways one can study the structure and dynamics of solids and liquids at the atomistic level using MD. In fact, a main reason why molecular simulation has become so widely used is the detailed information it can provide on the distribution of atoms and molecules in various states of matter, and the way they move about in response to thermal, mechanical, chemical, and biological excitations, natural or synthetic. Figure 2.3 is an early illustration comparing atomic trajectories in a typical crystal, liquid, and gas. Corresponding to the atomic structure of these systems, highly ordered

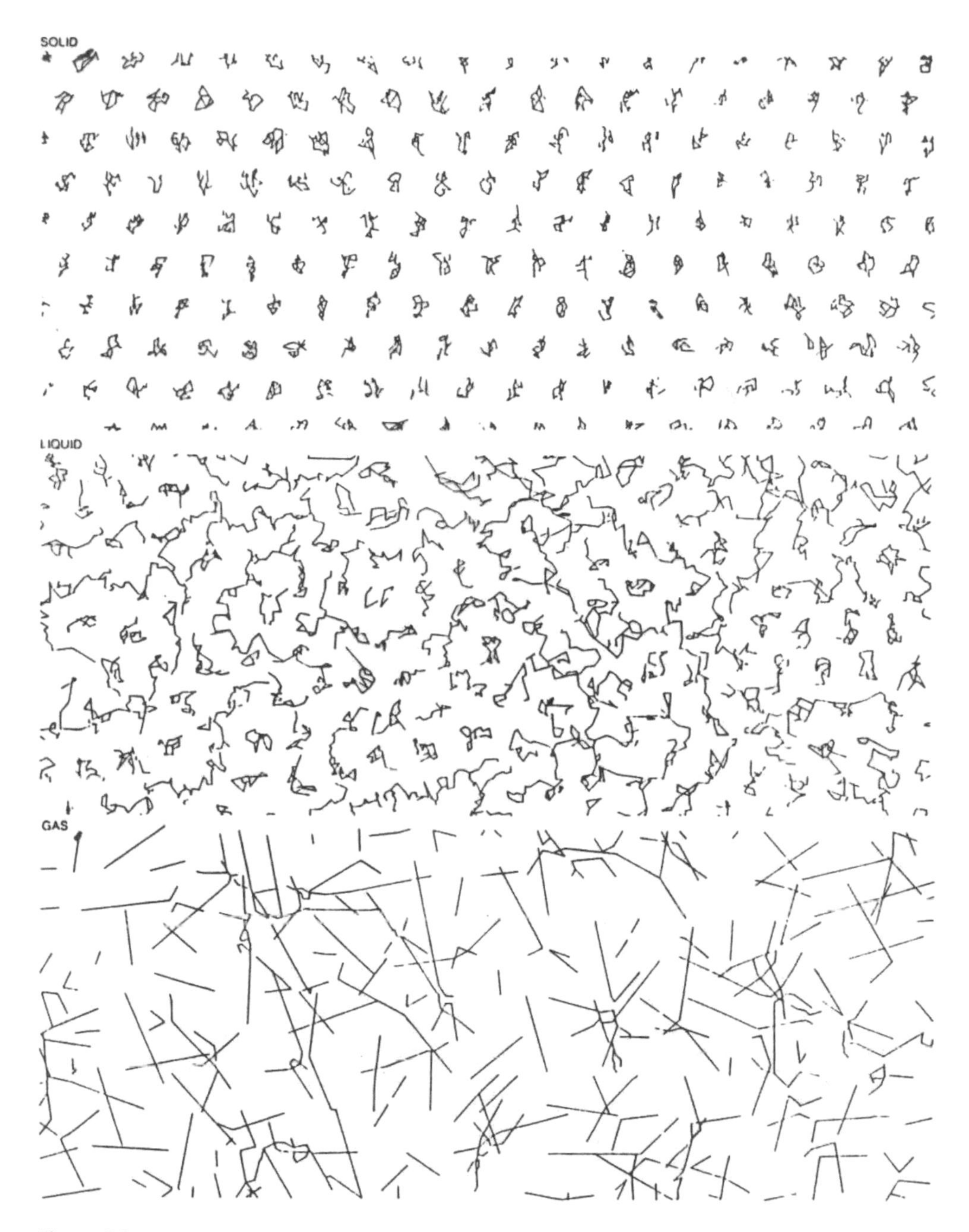

Figure 2.3
Atomic trajectories of a two-dimensional crystal, liquid, and gas simulated by molecular dynamics. Figure from (Barker and Henderson 1981).

in the solid state, quite disordered (but with some degree of local packing) in the liquid state, and uniformly random in the gaseous state, one can readily appreciate the respective particle motions being small-amplitude vibrations, diffusive movements over a local region, and long free flights interrupted by a collision every now and then. The basic challenge in simulation is how to quantify the detailed data seen in figure 2.3 in order to understand and then optimize the many physical properties of systems that can be studied, including the ones discussed in this work.

References

Allen, M. P., and D. J. Tildesley. 1987. *Computer Simulation of Liquids*. Oxford: Clarendon.

Barker, J. A., and D. Henderson. 1981. "The Fluid Phases of Matter." *Scientific American* 245 (5): 130–139.

Cai, W., J. Li, B. P. Uberuaga, and S. Yip. 2020. "Molecular Dynamics." In *Comprehensive Nuclear Materials*, 2nd ed., Vol. 1, edited by Rudy J.M. Konings, and Roger E. Stoller, 573–594. Oxford: Elsevier.

Further Reading

For an introduction to the MC simulation method mentioned in this essay but not discussed in any detail, see K. Binder, *Monte Carlo Methods in Statistical Physics*, 2nd ed, Springer (1986).

Yip, S. 1987. "Atomistic Simulations in Materials Science." In *Molecular Dynamics Simulation of Statistical-Mechanics Systems*, edited by G. Ciccotti and W. G. Hoover, 523. Amsterdam: North-Holland. For illustrative MD and MC results in early studies.

Haile, J. M. 1992. *Molecular Dynamics Simulation: Elementary Methods*. New York: Wiley. For a review, see Ian Johnston, "Molecular dynamic stimulation: elementary methods," *Computer in Physics* 7: 625 (1993).

Li, J. 2005. "Basic Molecular Dynamics." In *Handbook of Materials Modeling*, edited by S. Yip, 565. New York: Springer. Discussion of underlying techniques for students and nonexperts.

Gilmer, G., and S. Yip. 2005. "Basic Monte Carlo Models: Equilibrium and Kinetics." In *Handbook of Materials Modeling*, edited by S. Yip, 1931. Dordrecht: Springer. A primer on stochastic sampling using random numbers in contrast to deterministic sampling via solving Newton's equations in MD.

For a historical account, see G. Battemelli, G. Ciccotti, and P. Greco, *Computer Meets Theoretical Physics, The New Frontier of Molecular Simulation*, Springer (2020). An unconventional collaboration between a historian, physicist, and journalist.

Especially recommended for newcomers to molecular simulation is the LAMMPS simulation package, see A. P. Thompson et al., LAMMPS—A flexible simulation tool for particle-based materials modeling at the atomistic, meso, and continuum scales, *Computer Physics Communications* 271: 108171 (2022), and also the LAMMPS website, https://www.lammps.org/.

Following is an illustrative collection of early MD and MC studies, some with brief annotations.

Ortoleva, P., and S. Yip. 1976. "Computer molecular dynamics studies of a chemical instability." *The Journal of Chemical Physics* 65: 2045. Modeling the stability of a reactive fluid.

Kwok, T., P. S. Ho, S. Yip, R. W. Balluffi, P. D. Bristowe, and A. Brokman. 1981. "Evidence for vacancy mechanism for grain-boundary diffusion in Fe: a molecular dynamics study." *Physical Review Letters* 47: 1148. Mechanism of diffusion in a model bicrystal.

Chou, D. P., and S. Yip. 1982. "Computer molecular dynamics simulation of thermal ignition in a self-heating slab." *Combustion and Flame* 47: 215. Molecular simulation of thermal explosion.

Najafabadi, R., and S. Yip. 1983. "Observation of Bain transformation in alpha-Fe by Monte-Carlo simulation." *Scripta Metallurgica* 17: 1199. Phase diagram for bcc–fcc transformation at finite temperature.

Combs, J. A., and S. Yip. 1984. "Molecular dynamics study of kink diffusion." *Physical Review B: Condensed Matter and Materials Physics* 29: 438. Diffusion in a one-dimensional chain with thermal activation.

Ullo, J. J., and S. Yip. 1985. "Dynamical transition in a dense fluid approaching structural arrest." *Physical Review Letters* 54: 1509. Molecular simulation probing the onset of vitrification.

Anderson, J., J. J. Ullo, and S. Yip. 1987. "Molecular dynamics simulation of dielectric properties of water." *The Journal of Chemical Physics* 87: 1726. Quantitative determination of the dielectric constant.

Hsieh, H., and S. Yip. 1987. "Defect-induced crystal to amorphous transition in an atomistic model." *Physical Review Letters* 59: 2760. Crystal disordering without melting.

Limoge, Y., A. Rahman, H. Hsieh, and S. Yip. 1988. "Computer simulation studies of radiation-induced amorphization." *Journal of Non-Crystalline Solids* 99: 75.

Cheung, K. S., R. J. Harrison, and S. Yip. 1992. "Stress-induced Martensitic transition in a molecular dynamics model of alpha-iron." *Journal of Applied Physics* 71: 4009.

Chou, D. P., T. Lackner, and S. Yip. 1992. "Fluctuation effects in models of adiabatic explosion." *Journal of Statistical Physics* 69: 193.

Poon, T. W., P. Ho, F. F. Abraham, and S. Yip. 1992. "Ledge interactions and stress relaxations on Si(100) stepped surfaces." *Physical Review B: Condensed Matter and Materials Physics* 45: 3521. Intrinsic stability of stepped surfaces in semiconductors.

Yip, S. 1992. "Simulation studies of interfacial phenomena—melting, stress relaxation and fracture." In *Molecular Dynamics Simulations*, edited by F. Yonazawa, 221. Berlin: Springer-Verlag.

Wang, J., S.Yip, S. Phillpot, and D. Wolf. 1993. "Intrinsic response of crystals to pure dilatation." *Journal of Alloys and Compounds* 194: 407. Molecular details of stress-induced lattice instability.

Sonwalkar, N., and S. Yip. 1994. "Thermally induced fragmentation in an ice lattice." *The Journal of Chemical Physics* 100: 3747.

Li, J., Liao, D., and S. Yip. 1998. "Coupling continuum to molecular dynamics simulation: reflecting particle method and the field estimator." *Physical Review E* 57: 7259.

Li, J., Porter, L., and S. Yip. 1998. "Atomistic modeling of finite-temperature properties of crystalline beta-SiC II. Thermal conductivity and effects of point defects." *Journal of Nuclear Materials* 255: 139.

Kob, W. 1999. "Computer simulation of supercooled liquids and glasses." *Journal of Physics: Condensed Matter* 11: R85. Comprehensive review of advantages and drawbacks of MD simulation in the study of the dynamics of supercooled liquids with still quite relevant discussions of a currently active research topic.

Cai, W., M. de Koning, V. V. Bulatov, and S. Yip. 2000 "Minimizing boundary reflections in coupled-domain simulations." *Physical Review Letters* 85: 3213.

de Koning, M., W. Cai, A. Antonelli, and S. Yip. 2000. "Efficient free energy calculations by simulation of nonequilibrium processes." *Computing in Science and Engineering* 2: 88.

Yip, S., M. F. Sylvester, and A. S. Argon. 2000. "Atomistic investigation of segmental mobility in Atactic Poly(propylene)." *Computational and Theoretical Polymer Science* 10: 235.

Lin, X., J. Li, C. Foerst, and S. Yip. 2006. "Multiple self-localized soliton states in trans-polyacetylene." *Proceedings of the National Academy of Sciences of the United States of America* 103: 8943.

Silva, E. C. C. M., J. Li, D. Liao, S. Subramanian, T. Zhu, and S. Yip. 2006. "Atomic scale chemo-mechanics of silica: nanorod deformation and water reaction." *Journal of Computer-Aided Materials Design* 13: 135.

Trickey, S. B., S. Yip, H. Cheng, K. Runge, and P. A. Deymier. 2006. "A perspective on multi-scale simulation: toward understanding water-silica." *Journal of Computer-Aided Materials Design* 13: 1.

Zhu, W., D. E. Taylor, A. R. Al-Derzi, K. Runge, S. B. Trickey, Ju Li, Ting Zhu, and S. Yip. 2006. "Encoding electronic-structure information in potentials for multi-scale simulations: SiO_2." *Computational Materials Science* 38: 340.

Eapen, J., J. Li, and S. Yip. 2007. "Mechanism of thermal transport in dilute nanocolloids." *Physical Review Letters* 98: 028302.

Kaburaki, H., J. Li, S. Yip, and H. Kimizuka. 2007. "Dynamical thermal conductivity of Argon." *Journal of Applied Physics* 102: 043514.

Li, J., T. J. Lenosky, C. J. Först, and S. Yip. 2008. "Thermomechanical and mechanical stabilities of the oxide scale of ZrB_2 + SiC and oxygen transport mechanisms." *Journal of the American Ceramic Society* 91: 1475.

Qian, X., J. Li, L. Qi, C.-Z. Wang, T.-L. Chan, Y.-X. Yao, K.-M. Ho, and S. Yip. 2008. "Quasiatomic orbitals for ab initio tight-binding analysis." *Physical Review B: Condensed Matter and Materials Physics* 78: 245112.

Pellenq, R. J.-M., A. Kushima, R. Shahsavari, K. J. Van Vliet, M. J. Buehler, S. Yip, and F.-J. Ulm. 2009. "A realistic molecular model of cement hydrate." *Proceedings of the National Academy of Sciences of the United States of America* 106: 16102.

Masoero, E., E. Del Gado, R. J.-M. Pellenq, F.-J. Ulm, and S. Yip. 2012. "Nanostructure and nanomechanics of cement: polydisperse colloidal packing." *Physical Review Letters* 109: 155503.

Short, M. P., D. Hussey, B. K. Kendrick, T.M. Besmann, C.R. Stanek, and S.Yip. 2013. "Multiphysics modeling of porous CRUD deposits in nuclear reactors." *Journal of Nuclear Materials* 443: 579.

Del Gado, E., K. Ioannidou, E. Masoero, A. Baronnet, R.J.-M. Pellenq, F.-J. Ulm, and S. Yip. 2014. "A soft matter in construction—statistical physics approach to formation and mechanics of C-S-H gels in cement." *European Physical Journal Special Topics* 223: 2285.

Masoero, E., E. Del Gado, R. J.-M. Pellenq, S. Yip, and F.-J. Ulm. 2014. "Nanoscale mechanics of colloidal C-S-H gels." *Soft Matter* 10: 491.

Qomi, M. J. A., K. J. Krakowiak, M. Bauchy, K. L. Stewart, R. Shahsavari, D. Jagannathan, D. B. Brommer, A. Baronnet, M. J. Buehler, S. Yip, F.-J Ulm, K. J. Van Vliet, and R. J-.M. Pellenq. 2014. "Combinatorial molecular optimization of cement hydrates." *Nature Communications* 5: 4960.

Pinson, M. B., E. Masoero, P. A. Bonnaud, H. Manzano, Q. Ji, S. Yip, J. J. Thomas, M. Z. Bazant, K. J. Van Vliet, and H. M. Jennings. 2015. "Hysteresis from multiscale porosity: water sorption and shrinkage in cement paste." *Physical Review Applied* 3: 064009.

3 Particle Localization

Mode-coupling theory is a particular approximation in the theory of time correlations in nonequilibrium statistical mechanics. It is a significant extension of the Boltzmann transport equation, well known to be the foundation in the study of fluids from low density to around the critical point. For transport and fluctuation phenomena near the triple point or liquid densities, dynamical complexities of correlated binary collisions and collisions involving three or more particles are expected to dominate. These effects are mathematically difficult to analyze, even when appropriate extensions could be theoretically formulated. With the advent of neutron and light scattering experiments combined with the development of atomistic simulations, this was a compelling moment for the emerging interest in liquid-state dynamics in the 1960s.

The theoretical challenge confronting the community could be expressed in different ways; fundamentally, they all have the objective of deriving a tractable, molecular description of particle transport at liquid densities and arbitrary wavelengths and frequencies. The initial motivation of mode-coupling theory of liquids was to predict the structural relaxation and transport behavior over a wide range of frequencies and wavelengths. It was then realized that such a formulation could actually describe a particle localization mechanism involving a nonlinear feedback. In 1984, it was further realized that within self-consistent mode-coupling formulation one could develop a description that predicts an idealized liquid to glass transition. This result, the prospect of obtaining a molecular description of the glass transition phenomenon, was met with considerable excitement in the community. While the mode-coupling approximation could account for the nonlinear response of feedback dynamics, further extensions would be needed to account for the activated kinetics of potential energy barrier hopping.

Introduction

In the 1980s, there was much interest in the theoretical description of density fluctuations in fluids that have been either cooled or compressed beyond the freezing point. Because these are in highly nonequilibrium states, the dynamics of density fluctuations in these systems could be significantly different. A reason for excitement at the time was the discovery that a certain self-consistent mode-coupling approximation could lead to a tractable theoretical description of a freezing transition, where the system becomes nonergodic (Leutheusser 1984; Bengtzelius, Götze, and Sjolander 1984).

The aim of this essay is to briefly explain the essence of mode-coupling theory. It is intended to be a qualitative look at a fundamental area of theoretical research in non-equilibrium statistical mechanics (Götze 2007; Yip 1989).

Mode-Coupling Approximations (Schematic)

In essay 1, we posed the problem of density fluctuations in an atomic fluid in equilibrium at arbitrary density n and temperature T (Boon and Yip 1980). To explain the basic physical idea of mode-coupling, we consider analyzing the problem schematically. Recall the intermediate scattering function $F(k, t)$, the Fourier transform of the space-time density correlation function $G(r, t)$ and define $\varphi(z)$ as its Laplace transform

$$\varphi(z) = i \int_0^\infty dt\, e^{izt} F(t) \tag{3.1}$$

The wavenumber dependence in F will be suppressed for simplicity.

There are two ways to calculate $\varphi(z)$. One is to derive a kinetic equation for the phase-space density correlation function whose momentum integral gives $\varphi(z)$ (Leutheusser 1984; Bengtzelius, Götze, and Sjolander 1984). Alternatively, one can adopt the memory function approach by introducing the memory function $K(z)$ heuristically (Boon and Yip 1980; Yip 1989)

$$-\varphi(z)^{-1} = z + K(z) \tag{3.2}$$

and further express $K(z)$ in terms of its *own* memory function $M(z)$

$$-\Omega^2 K(z)^{-1} = z + M(z) \tag{3.3}$$

where $\Omega(k) = kv_0/\sqrt{S(k)}$ is a characteristic frequency of the fluid, v_0 being the thermal speed and $S(k)$ the static structure factor with its maximum at $k = k_0$.

In the self-consistent mode-coupling formulation an approximate expression for $M(z)$ is postulated by writing $M(t)$ as the sum of two components, a short-time relaxation due to collisions among *individual* particles and a long-time decay associated with *collective* behavior such as motions of clusters of particles. The first component represents the Enskog approximation where only uncorrelated binary collisions are considered. This part of $M(t)$ is well known. The long-time decay represents coupling among different hydrodynamic modes such as density, current, and temperature fluctuations; this is the part that makes the theory physically reasonable at high densities or low temperatures.

We are concerned with two closely related but distinct mode-coupling approximations, which schematically can be written as (Yip 1989)

$$M(z) \simeq M_0(z) + m(z) \tag{3.4}$$

and

$$M(z) \simeq M_0(z) + \frac{m(z)}{1 - \Delta(z)m(z)} \tag{3.5}$$

with

$$M_0(t) = \nu\delta(t) \tag{3.6}$$

$$m(t) = \lambda_1 F(t) + \lambda_2 F^2(t) \tag{3.7}$$

$$\Delta(t) = \lambda_3 F(t)\, J(t) \tag{3.8}$$

The memory function $M_0(t)$ is the Enskog approximation, with ν being the collision frequency. To focus attention on the collective behavior described by $m(z)$, we have taken the relaxation described in $M_0(t)$ to be instantaneous, thereby ignoring the dynamical features on the timescale of the duration of binary collisions. The relaxation function $m(t)$ is the simplest mode-coupling treatment of collective effects; in the scheme considered the coupling involves only linear and quadratic density modes. The density- and temperature-dependent coupling coefficients λ_1 and λ_2 will be treated as constants here for simplicity, but if needed, explicit expressions for them can be given in terms of the interatomic potential function and the two- and three-particle radial distribution functions. The function $\Delta(t)$ arises when coupling between the density and longitudinal current fluctuations are taken into consideration, with $J(t)$ being the longitudinal current correlation function and λ_3 another coupling coefficient. The presence of Δ is the distinction between the two approximations, equations (3.4) and (3.5).

The remarkable properties of the self-consistent mode-coupling formalism were first demonstrated using equation (3.4), and with $\lambda_1 = 0$ in equation (3.7) (Leutheusser 1984; Bengtzelius, Götze, and Sjolander 1984). It was shown that the memory function contribution $m(t)$ led to an ergodic-to-nonergodic transition at a critical density n_c or temperature T_c, at which point the diffusivity vanishes and the shear viscosity diverges with the appearance of a nonzero shear modulus (Bengtzelius, Götze, and Sjolander 1984). Further studies were reported (Kirkpatrick 1985; Götze 1984; Frederickson 1988). An equivalent formulation in terms of a nonlinear fluctuating hydrodynamic theory

(Das et al. 1985) confirmed this relatively simple hypothesis was capable of giving a rich spectrum of nonlinear relaxation and transport properties. We will henceforth refer to equations (3.4) and (3.7), in which $\lambda_1 = 0$, as the LBGS approximation.

The onset of nonergodicity signifies the freezing in of some of the structural degrees of freedom in the fluid, a transition that can be regarded as an ideal liquid-to-glass transition since the model contains no mechanism for crystallization. The origin of this transition is purely dynamical; it can be traced to a nonlinear feedback mechanism that leads to particle localization (Götze, Leutheusser, and Yip 1981a; Götze, Leutheusser, and Yip 1981b; Götze, Leutheusser, and Yip 1982; Geszti 1983). Notice in the present formulation that structural arrest occurs when $F(t \to \infty) = f$ is no longer zero, or, in view of equation (3.1), $\varphi(z)$ has a zero-frequency pole, $\varphi(z) \sim 1/z$. This can come about when $m(z) \sim 1/z$, so that $M(z)$ becomes likewise singular.

The second approximation, equation (3.5), results when one takes into account the coupling between density and current fluctuations in deriving the mode-coupling contribution to $M(z)$ (Das and Mazenko 1986; Götze and Sjögren 1987a; Götze and Sjögren 1987b). Thus, it can be regarded as an extension of the LBGS approximation. The effect of this coupling is seen to be a renormalization of the memory function such that the singular behavior of $M(z)$ is cutoff; in other words, when $\Delta \neq 0$, $M(z)$ no longer diverges like $1/z$ for small z even though $m(z) \sim 1/z$, so the system always remains ergodic. On the other hand, it does not mean that the two approximations, equations (3.4) and (3.5), necessarily have to give different results in the intermediate-time domain where significant slowing down in the relaxation occurs due to collective effects. The complexity of equation (3.5) has thus far precluded much analytical analysis of its properties; there exist limited numerical results (Das and Mazenko 1986; Das 1987; Götze and Sjögren 1988) that suggest that in the time domains appropriate to neutron and light scattering measurements and MD simulations, the two descriptions may give quite similar decay behavior of $F(t)$.

Coupling of Density Fluctuations

The basic hypothesis of the mode-coupling formulation is that the part of $M(t)$ representing the collective behavior, or the effects of correlated collisions in the formalism of renormalized kinetic theory (Mazenko 1974), can be expressed as products of two hydrodynamic modes. This hypothesis has not been validated by estimating the leading corrections. Indeed, validation of various mode-coupling approximations generally has to rely on the comparison of model results with experimental or computer simulation data.

Consider first the validity of the LBGS predictions. At densities below a critical density n_c, the self-diffusion coefficient is predicted to have a power-law density dependence, $D \sim (n_c - n)^{\alpha}$, with exponent $\alpha \simeq 1.76$ and correspondingly the reciprocal of the shear viscosity η behaves in the same way. There exist diffusivity data for the supercooled liquid methyl-cyclohexane and from simulation results on hard-sphere and Lennard-Jones fluids that can be fitted to this power law (Bengtzelius, Götze, and Sjolander 1984); also, there are several "fragile" liquids whose viscosity in the supercooled region follows a temperature variation $\eta \sim (T - T_o)^{-2}$, where T_o is a temperature distinctly greater than the glass transition point T_g (Taborek, Kleiman, and Bishop 1986). Thus, one may conclude that the density and temperature variations of the transport coefficients of simple fluids in the dense liquid and super-cooled regimes can be predicted reasonably well by the LBGS approximation. This in itself is already a significant improvement over the Enskog theory, which is generally valid up to about half the liquid density and is known to fail at around the triple-point density (Alder and Wainwright 1970).

Numerical calculations of the density correlation function $F(k, t)$ have been performed for the LBGS model (Bengtzelius 1986), and the results at various densities directly compared with MD simulation data on a Lennard-Jones system (Ullo and Yip 1985; Ullo and Yip 1989). It was found that while both model and simulation results for the decay of $F(t)$ show a characteristic slowing down due to structural relaxation, the densities at which similar behavior sets in do not correspond very well; specifically, the model seems to predict the onset of structural arrest at a lower density compared to the simulation results.

The more general approximation of equations (3.4) and (3.7) with nonzero λ_1 and λ_2 has also been analyzed in some detail (Götze and Sjögren 1987a; Götze and Sjögren 1987b; Buchalla et al. 1988). It was shown that with $\lambda_1 \neq 0$ the model exhibits scaling and stretching behavior characteristic of the relaxation behavior typical of systems near the liquid-glass transition (Wong and Angell 1976). The property of scaling means $F(t)$ can be written in the form of $F(t) = \Lambda(t/\tau)$, where Λ is a universal or master function and relaxation time τ is a parameter which accounts for all the temperature dependence. Stretching, on the other hand, manifests in a nonexponential decay of $F(t)$, $F(t) = F_0 \exp(-(t/\tau)^{\beta})$, with $0 < \beta < 1$. It is rather remarkable that the approximation of equations (3.4) and (3.7) can predict such behavior (Götze 1984; Götze and Sjögren 1987a; Götze and Sjögren 1987b) which, moreover, have been found to have experimental correspondence in inelastic neutron scattering studies (Mezei, Knaak, and Farago 1987). If, instead of equation (3.7), one writes $m(t) = \Gamma(F(t))$, where Γ is a mode-coupling functional, a polynomial with nonnegative coefficients, then one can even discuss the phenomenon of β relaxation, which has been observed in dielectric loss spectra (Johari and Goldstein 1970; Johari and Goldstein 1971) and is also characterized by a broad distribution of relaxation rates.

The present assessment of the validity of the mode-coupling approximation therefore finds the correct prediction of several fundamental features of relaxation and transport behavior that have been observed in supercooled liquids. The harder question of whether the ideal glass transition predicted by the LBGS model has anything to do with the glass transition observed experimentally remains to be addressed. A major difficulty is that the definition of the latter depends on the physical property one is considering, and different changes in structural, thermodynamic, or transport properties have been used to characterize the glass transition in various measurements. Since the ideal glass transition is purely dynamical in origin, it seems natural to examine its relation to the behavior of a transport coefficient, the shear viscosity coefficient η. In this context it is a matter of empirical practice to define a glass transition temperature T_g as that temperature at which η has the value of 10^{13} P (see figure 11.1 for a plot of the viscosity η of supercooled glass-forming liquids). The question then becomes whether the transition predicted by the LBGS model is capable of describing the magnitude and the temperature variation of η in the region near T_g. In other words, is it reasonable to identify T_c with T_g?

When the question is posed in this manner it follows from the theoretical side the difference between the two approximations, equations (3.4) and (3.5), can be quite important; moreover, one needs to obtain numerical results for η that can be compared with experiments or computer simulation data. From the experimental side, it has been pointed out that the shear viscosity coefficients of a number of supercooled liquids all show a typical behavior of a power-law temperature variation, with exponent approximately 2, in the regime where η has increased to about 10 P, and that at lower temperatures η increases much more rapidly (Taborek, Kleiman, and Bishop 1986). By extrapolating the data in the low-η region to infinite viscosity one obtains a characteristic temperature which the investigators have denoted as T_o. The significance of T_o is that it is the "transition temperature" in the supercooled region separating the two characteristic behaviors of the shear viscosity. Thus, η behaves like the prediction of LBGS for $T > T_o$, while in the region of $T < T_o$, the data show an increase by several orders of magnitude until one reaches T_g. Furthermore, the latter increase occurs over a considerable range of temperature, so T_o is well separated from T_g, and in some cases $T_o \sim 2T_g$.

To our knowledge, one numerical calculation of η using an approximation similar to equation (3.5) has been reported (Das and Mazenko 1986; Das 1987). From the rather limited results obtained, one sees that relative to its value in the normal liquid state, η in the supercooled region can show increases of a factor of 10–100, but not several orders of magnitude. The implication is that either equation (3.4) or equation (3.5) is unable to describe well the sharp increase of η in the temperature region below T_o.

This conclusion actually could be rationalized by noting that activated-state dynamics is not taken into account in the LBGS model. For the extended model, equation (3.5), one may regard the presence of Δ as the one-phonon contribution to phonon-assisted hopping processes (Götze 2007). This is an issue that remains to be clarified.

The T_x Problem

If it is correct that the present mode-coupling approximations do not apply to the glass transition phenomenon observed in the laboratory, one may still ask what is the significance of the temperature region near T_0? The behavior of η suggests that T_0 delineates a crossover region where the particle motions apparently change from continuous fluid-like displacements to barrier hopping between potential energy minima. Such a picture of viscous flow was considered some thirty years ago (Goldstein 1969). Within this scenario, one may think of a broad dynamical transition that signals the onset of activation barriers on the potential energy landscape (PEL) (see also essays 8 and 11 for the relevance of the PEL concept). It has been suggested to label the crossover transition by a temperature T_x instead of T_0 in order to avoid confusion with a parameter in the well-known empirical expression for η or D (Angell 1988).

There is evidence in various forms pointing to the existence of a crossover transition. An MD study of compressed Lennard-Jones fluids has revealed such a behavior, the transition being characterized by a decrease in the compressibility, a change in the density variation of D, and the onset of a slowly decaying component in $F(t)$, all occurring at a density well below that required to bring the diffusivity down to typical values for glasses (Ullo and Yip 1989). Another MD study of local stress fluctuations in quenched liquids showed the onset of spatial correlation effects at a characteristic temperature considerably above T_g (Chen, Egami, and Vitek 1988). In recent studies of spin-glass models, it has been found that two transition temperatures can be identified, a dynamical transition at higher temperature that is associated with the appearance of barriers in the local energy surface and an equilibrium transition at a lower temperature where the configurational entropy vanishes (Kirkpatrick and Thirumalai 1987).

Diffusion and Localization in a Random, Static Field

Among the various models in statistical physics that have been developed to study particle diffusion in matter is a class called the Lorentz models (Hauge 1974). In these highly idealized models, the classical point particles move in a medium of *randomly distributed stationary scattering sites* and they interact only with these sites and not with each other. Different Lorentz models are distinguished by the shape of the scattering sites. One can also specify whether the sites, which can have finite extents, are allowed to overlap, and the spatial dimension. Even though the medium is static, the Lorentz models are of conceptual interest because they exhibit nontrivial correlation effects that are also characteristic of diffusion in normal liquids. In a density expansion of the diffusion coefficient D, one encounters divergences of the same kind and resummations leading to nonanalytic terms (Weyland and van Leeuwen 1968). It is also known from kinetic theory analysis of ring collisions the velocity autocorrelation function decays nonexponentially at long times, an intriguing behavior first established by MD simulations (Ernst and Weyland 1971).

Theoretical results on the hard-sphere Lorentz model have been checked by MD simulation to reveal nonanalytic behavior in the density variation of D (Bruin 1972; Bruin 1974) and the power law decay of the velocity autocorrelation function in two dimensions (Bruin 1972; Bruin 1974; Alder and Alley 1978; Lewis and Tjon 1978). Additionally, simulation results on overlapping hard discs have established the critical density at which D vanishes, and the detailed behavior of the velocity autocorrelation function above and below the critical region (Alder and Alley 1978).

Here, we briefly recall the results of mode-coupling theory applied to the two-dimensional (2D) spherical Lorentz model with overlap, commenting on the relevant findings for the diffusion coefficient, the velocity autocorrelation, and the self-correlation function (Götze, Leutheusser, and Yip 1981a; Götze, Leutheusser, and Yip 1981b; Götze, Leutheusser, and Yip 1982). For particles moving with constant thermal speed v_0, the relation between the density correlation $\phi(q, z)$ and its memory function $M(z)$ associated with a kinetic equation is

$$\phi(q,z) = \frac{\phi^{(0)}(q, z + M(z))}{1 + M(z)\phi^{(0)}(q, z + M(z))} \tag{3.9}$$

where $\phi^{(0)}(q, z)$ is the free-particle density-correlation function

$$\phi^{(0)} = -\frac{1}{(z^2 - q^2 v_0^2)^{1/2}} \tag{3.10}$$

Analysis of $M(z)$ begins by separating it into two contributions

$$M(z) = i\nu + m(z) \tag{3.11}$$

Here $\nu = 8n\sigma v_0/3$ is the binary-collision frequency, where n is the density of static hard-disk scatterers whose radius is σ. The first term represents the effects of uncorrelated binary collisions as treated in conventional transport theory, whereas $m(z)$ represents the contributions from correlated collisions.

As indicated in our earlier discussion, the essence of mode-coupling approximation is to express $m(z)$ back in terms of the time correlation of interest, $\phi(q, z)$, or equivalently $M(z)$

$$m(z) = -\frac{\pi n^*}{2} M(z) \times \int_0^\infty \frac{dk}{k} \frac{F(k \cdot \sigma, [z + M(z)]/kv_0)^2}{1 + M(z)\phi^{(0)}(k, z + M(z))} \tag{3.12}$$

where $n^* = n\sigma^2$ is the reduced density, and

$$F(k, x) = J_0'(k)\phi_o(x) - i\frac{16}{3\pi} J_1'(k)\phi_1(x) - J_2'(k)\phi_2(x) \ldots \tag{3.13}$$

$$\phi_0(x) = -(x^2 - 1)^{-1/2} \tag{3.14a}$$

$$\phi_1(x) = 1 + x\phi_0(x) \tag{3.14b}$$

$$\phi_2(x) = 2x\phi_1(x) - \phi_0(x) \tag{3.14c}$$

The $J_n'(k)$ are the Bessel function derivatives of index n. Combining the preceding formulas one is confronted with a transcendental equation for $M(z)$, which could be solved by iteration starting at large frequencies with $M(z) \cong iv$. It follows from equations (3.9) and (3.10) that the velocity-autocorrelation function is given by

$$K(z) = -\frac{v_0^2/2}{z + M(z)} \tag{3.15}$$

along with the self-diffusion coefficient

$$D = v_0^2/2M''(\omega = 0) \tag{3.16}$$

where $M''(\omega)$ denotes the imaginary part of $M(z = \omega + i0)$.

In figure 3.1, two sets of computer simulation results (Bruin 1972; Alder and Alley 1978) on the density variation of the diffusion coefficient are compared with the mode-coupling results of equation (3.16). Both the diffusion coefficient and the density are shown in reduced units, $D/D^{(0)}$ and n/n_c, where $D^{(0)}$ is the low-density value of the diffusion coefficient and n_c is the critical density for the model system. Notice that there is a qualitative distinction between the Lorentz model and the hard-sphere fluid where one has persistence of velocity correlations that results in $D > D^{(0)}$, as well as the cage effect

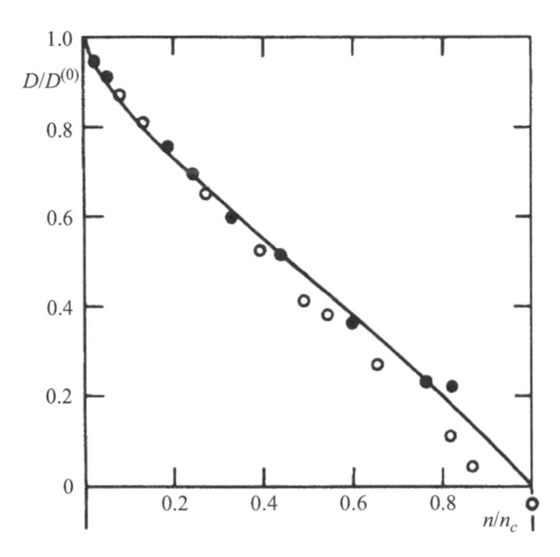

Figure 3.1

Variation of reduced diffusion coefficient with reduced density for the Lorentz model of overlapping hard discs, mode-coupling model calculation (curve [Götze 1982]), and computer-simulation results (closed circles [Bruin 1972], open circles [Alder and Alley 1978]). Reprinted figure with permission from (Götze, Leutheusser, and Yip 1982). Copyright 1982 by American Physical Society.

setting in at high densities and thereby causing $D/D^{(0)}$ to decrease sharply. By contrast, the static medium in the Lorentz system eliminates the possibility of vortex modes and hence the enhancement effect, but the stationary nature of the scattering sites still allows spatial correlations to be built up through repeated collisions to generate a cage effect.

The onset of particle localization in molecular simulations is thus well captured in an explicit, dynamical description of a feedback effect by expressing the current-relaxation kernel $M(z)$ as the sum of contributions from uncorrelated binary collisions and from correlated collisions. While both are explicitly proportional to n, the latter also varies inversely with D. Since $M(z \to 0)$ plays the role of an inverse diffusion coefficient, the fact that it contains a contribution varying like $1/D$ is indicative of a feedback effect that allows the correlated collisions to dominate over uncorrelated binary collisions. Schematically, one can express the low-frequency limit of $M(z)$ in the form $1/D = 1/D^{(0)} + (n/n_c)/D$. At low densities, $n/n_c \ll 1$, D is governed mainly by uncorrelated binary collisions so that $D \sim D^{(0)}$. As n increases, correlated collision effects become significant not only because of the density factor but also $D/D^{(0)}$ can be appreciably smaller than unity. As one approaches the critical density, D becomes very small as the effective collision frequency becomes very large. This then is the dynamical picture of the "cage effect." Figure 3.1 demonstrates the mode-coupling theory accounts for this behavior reasonably well.

The analysis of the Lorentz model has been extended to the velocity autocorrelation function and the van Hove self-correlation function (Götze, Leutheusser, and Yip 1982). We will not discuss the specific details any further except to note that localization phenomena are of general interest in the understanding of molecular mechanisms. We will encounter several instances where they play key roles in dynamical complexities, such as in crystal plasticity (part III) and amorphous rheology (part V).

Outlook on Nonlinear Feedback Mechanisms

In closing our commentary on the mode-coupling formalism, we offer two observations. First is that an appropriate assessment of this theoretical analysis should be carried out in the moderately supercooled temperature regime T_x and not in the deeply supercooled regime near T_g. The extended model, equation (3.13), should be studied to see if the effects of the cutoff of the nonergodic transition are numerically significant; in this respect, analysis and comparison with MD data on $F(t)$ will give further insights.

The other observation is the reminder that the mode-coupling approximation was initially developed to treat the collective aspects of collisional dynamics that could not be taken into account through the Enskog approximation in kinetic theory. In particular, the approximation of coupling to density and current fluctuations was not invoked heuristically only for the purpose of describing the liquid–glass transition. One should not expect such an approach could lead to a complete description of the liquid-to-glass transition without revision and extension. In view of the analysis of viscoelastic behavior in dense hard-sphere fluids (Leutheusser 1982a; Leutheusser 1982b) and diffusion-localization transition in the Lorentz model (Götze, Leutheusser, and Yip 1981a; Götze, Leutheusser, and Yip 1981b; Götze, Leutheusser, and Yip 1982), its intrinsic merit for describing the dynamical relaxation behavior of supercooled liquids and glasses is noteworthy. Further developments will be discussed in parts IV and V.

References

Alder, B. J., and W. E. Alley. 1978. "Long-time correlation effects on displacement distributions." *Journal of Statistical Physics* 19: 341.

Alder, B. J., and T. E. Wainwright. 1970. "Decay of the velocity autocorrelation function." *Physical Review A* 1: 18.

Angell, C. A. 1988. "Decay of the velocity autocorrelation function." *Journal of Physical and Chemistry of Solids* 49: 863.

Bengtzelius, U. 1986. "Dynamics of a Lennard-Jones system close to the glass transition." *Physical Review A* 34: 5059.

Bengtzelius, U., W. Götze, and A. Sjolander. 1984. "Dynamics of supercooled liquids and the glass transition." *Journal of Physics C: Solid State Physics.* 17: 5915.

Boon, J-P., and S. Yip. 1980. *Molecular Hydrodynamics.* New York: McGraw-Hill.

Bruin, C. 1972. "Logarithmic terms in the diffusion coefficient for the Lorentz gas." *Physical Review Letters* 29: 1670.

Bruin, C. 1974. "A computer experiment on diffusion in the Lorentz gas." *Physica* (Utrecht) 72: 261.

Buchalla, G., U. Dersch, W. Götze, and L. Sjögren. 1988. "Alpha-and beta-relaxation for single-particle motion near the glass transition." *Journal of Physics C* 21: 4239.

Chen, S.-P., T. Egami, and V. Vitek. 1988. "Local fluctuations and ordering in liquid and amorphous metals." *Physical Review B* 37: 2240.

Das, S. P. 1987. "Effects of structure on the liquid-glass transition." *Physical Review A* 36: 211.

Das, S. P., and G. F. Mazenko. 1986. "Fluctuating nonlinear hydrodynamics and the liquid-glass transition." *Physical Review A* 34: 2265.

Das, S. P., G. F. Mazenko, S. Ramaswamy, and J. J. Toner. 1985. "Hydrodynamic theory of the glass transition." *Physical Review Letters* 54: 118.

Ernst, M. H., and A. Weyland. 1971. "Long time behaviour of the velocity auto-correlation function in a Lorentz gas." *Physics Letters* 34A: 39.

Frederickson, G. 1988. "Recent developments in dynamical theories of the liquid-glass transition." *Annual Review of Physical Chemistry* 39: 149.

Geszti, T. 1983. "Pre-vitrification by viscosity feedback." *Journal of Physics C* 16: 5805.

Goldstein, M. 1969. "Viscous liquids and the glass transition: a potential energy barrier picture." *Journal of Chemistry and Physics* 51: 3728.

Götze, W. 1984. "Some aspects of phase transitions described by the self consistent current relaxation theory." *Zeitschrift für Physik B* 56: 139.

Götze, W. 2007. *Complex Dynamics of Glass Forming Liquids.* Oxford: Oxford Univ. Press.

Götze, W., E. Leutheusser, and S. Yip. 1981a. "Dynamical theory of diffusion and localization in a random, static field." *Physical Review A* 23: 2634.

Götze, W., E. Leutheusser, and S. Yip. 1981b. "Correlation functions of the hard-sphere Lorentz model." *Physical Review A* 24: 1008.

Götze, W., E. Leutheusser, and S. Yip. 1982. "Diffusion and localization in the two-dimensional Lorentz model." *Physical Review A* 25: 533.

Götze, W., and L. Sjögren. 1987a. "The glass transition singularity." *Zeitschrift für Physik B* 65: 415.

Götze, W., and L. Sjögren. 1987b. "α-relaxation near the liquid-glass transition." *Journal of Physics C* 20: 879.

Götze, W., and L. Sjögren. 1988. "Scaling properties in supercooled liquids near the glass transition." *Journal of Physics C* 21: 3407.

Hauge, E. H. 1974. "What can one learn from Lorentz models?" In *Transport Phenomena. Lecture Notes in Physics*, edited by G. Kirczenow and J. Marro, Vol. 31, 337. Berlin: Springer.

Johari, G. P., and M. Goldstein. 1970. "Viscous liquids and the glass transition. II. Secondary relaxations in glasses of rigid molecules." *Journal of Chemistry and Physics* 53: 2372.

Johari, G. P., and M. Goldstein. 1971. "Viscous liquids and the glass transition. III. Secondary relaxations in aliphatic alcohols and other nonrigid molecules." *Journal of Chemistry and Physics* 55: 4245.

Kirkpatrick, T. R. 1985. "Mode-coupling theory of the glass transition." *Physical Review A* 31: 939.

Kirkpatrick, T. R., and D. Thirumalai. 1987. "Mean-field soft-spin Potts glass model: statics and dynamics." *Physical Review B* 37: 5342.

Leutheusser, E. 1982a. "Dynamics of a classical hard-sphere gas I. Formal theory." *Journal of Physics C: Solid State Physics* 15: 2801.

Leutheusser, E. 1982b. "Dynamics of a classical hard-sphere gas. II. Numerical results." *Journal of Physics C: Solid State Physics* 15: 2827.

Leutheusser, E. 1984. "Dynamical model of the liquid-glass transition." *Physical Review A* 29: 2765.

Lewis, J. C., and J. A. Tjon. 1978. "Evidence for slowly-decaying tails in the velocity autocorrelation function of a two-dimensional Lorentz gas." *Physics Letters* 66A: 349.

Mazenko, G. F. 1974. "Fully renormalized kinetic theory. III. Density fluctuations." *Physical Review A* 9: 360.

Mezei, F., W. Knaak, and B. Farago. 1987. "Neutron spin echo study of dynamic correlations near liquid-glass transition." *Physical Review Letters* 58: 571.

Taborek, T., R. N. Kleiman, and D. J. Bishop. 1986. "Power-law behavior in the viscosity of supercooled liquids." *Physical Review B* 34: 1835.

Ullo, J. J., and S. Yip. 1985. "Dynamical transition in a dense fluid approaching structural arrest." *Physical Review Letters* 54: 1509.

Ullo, J. J., and S. Yip. 1989. "Dynamical correlations in dense metastable fluids." *Physical Review A* 39: 5877.

Weyland, A., and J. M. J. van Leeuwen. 1968. "Non-analytic density behaviour of the diffusion coefficient of a Lorentz gas: II. Renormalization of the divergencies." *Physica* (Utrecht) 38: 35.

Wong, J., and C. A. Angell. 1976. *Glass Structure by Spectroscopy*. New York: Marcel Dekker.

Yip, S. 1989. "Commentary on the self-consistent mode-coupling approximation." *Journal of Statistics and Physics* 57: 665.

Further Reading

This essay is based primarily on the following:

Yip, S. 1989. "Commentary on the self-consistent mode-coupling approximation." *Journal of Statistics and Physics* 57: 665.

Götze, W., E. Leutheusser, and S. Yip. 1981. "Dynamical theory of diffusion and localization in a random, static field." *Physical Review A* 23: 2634.

Götze, W., E. Leutheusser, and S. Yip. 1981. "Correlation function of the hard-sphere Lorentz model." *Physical Review A* 24: 1008.

Götze, W., E. Leutheusser, and S. Yip. 1982. "Diffusion and localization in the two-dimensional Lorentz model." *Physical Review A* 25: 533.

Ullo, J. J., and S. Yip. 1985. "Dynamical transition in a dense fluid approaching structural arrest." *Physical Review Letters* 54: 1509. Early MD results inspired by the development of a mode-coupling theory describing an ideal liquid-to-glass transition.

Readers interested in the early developments of Mode Coupling Theory (MCT) will find useful a special issue of *Transport Theory and Statistical Physics*, Volume 24, Number 6-8 (1995), devoted to Relaxation Kinetics in Supercooled Liquids—Mode-Coupling Theory and Its Experimental Tests (S. Yip, ed., New York: Marcel Dekker, 1995).

Additional references for readers interested in the later developments of MCT and applications:

Fuchs, M., and M. Cates. 2002. "Theory of nonlinear rheology and yielding of dense colloidal suspensions." *Physical Review Letters* 89: 248304. A first-principles approach to nonlinear flow of dense suspensions based on an extended mode-coupling theory formalism is shown to capture shear thinning and dynamical yielding.

Shi, Y., and M. L. Falk. 2005. "Strain localization and percolation of stable structure in amorphous solids." *Physical Review Letters* 95: 095502. MD simulation study of the effect of structural relaxation prior to mechanical testing, with more rapidly quenched initial structures undergoing more localization.

Brader, J. M., T. Voigtmann, M. E. Cates, and M. Fuchs. 2007. "Dense colloidal suspensions under Time-Dependent Shear." *Physical Review Letters* 98: 058301. Extended mode-coupling analysis of driven colloidal suspensions giving a time-dependent description of rheological response far from equilibrium.

Chaudhuri, P., L. Berthier, and W. Kob. 2007. "Universal nature of particle displacements close to glass and jamming transitions." *Physical Review Letters* 99: 060604. Distributions of single-particle displacement show the coexistence of slow and fast particles as a signature of dynamical heterogeneities.

Biroli, G. 2007. "A new kind of phase transition?" *Nature Physics* 3: 222. Jamming in three-dimensional binary mixture of glass-forming particles at high densities and slowly driven externally are identified as hallmarks of dynamical heterogeneities.

Logendijk, A., B. van Tiggelen, and D. S. Wiersma. 2009. "Fifty years of Anderson localization." *Physics Today* 24. What began as a prediction about electron diffusion has spawned a rich variety of theories and experiments on the nature of the metal-insulator transition and the behavior of waves—from electromagnetic to seismic—in complex materials.

Brader, J. M., T. Voigtmann, M. Fuchs, R. G. Larson, and M. E. Cates. 2009. "Glass rheology: from mode-coupling theory to a dynamical yield criterion." *Proceedings of the National Academy of Science* 106: 15186. A schematic (single mode) model describing the dynamical yield surface for a class of rheological flows.

Fuchs, M., and M. E. Cates. 2009. "Integration through transients for Brownian particle under steady shear." *Journal of Physics and Condensed Matter* 17: 51681.

Mallemace, F., C. Branca, C. Corsaro, N. Leone, J. Spooren, S.-H. Chen, and H. E. Stanley. 2010. "Transport properties of glass forming liquids suggest that dynamical crossover temperature is as important as the glass transition temperature." *Proceedings of the National Academy of Science* 107: 22457. Experimental data on shear viscosity and self-diffusion show a characteristic crossover temperature T_x from Arrhenius to super-Arrhenius behavior, below which the Stokes–Einstein relation breaks down.

Filoche, M., and M. Mayboroda. 2012. "Universal mechanism for Anderson and weak localization." *Proceedings of the National Academy of Science* 109: 14761. Localization of stationary waves in mechanical, acoustical, optical and quantum systems associated with an inhomogeneous medium, complex geometry, or quenched disorder manifested by partitions into weakly coupled subregions.

Das, S. P. 2011. *Statistical Physics of Liquids at Freezing and Beyond.* Cambridge Univ. Press. This is a treatise on the mode-coupling and fluctuating hydrodynamics formalisms.

Gruber, M., G. C. Abade, A. M. Puertas, and M. Fuchs. 2016. "Active microrheology in a colloidal glass." *Physical Review E* 94: 042602. Implementation of extended mode-coupling theory describing a driven probe particle in a colloidal glass of hard spheres, comparing calculations with simulations.

Maier, M., A. Zippelius, and M. Fuchs. 2017. "Emergence of long-ranged stress correlations at the liquid to glass transition." *Physical Review Letters* 119: 265701.

II Melting Scenarios

4 Elastic-Melting Instability

The criteria formulated by M. Born for the onset of melting in 1939 and lattice stability in 1940 are reexamined on the basis of elastic instabilities in the presence of an external stress. With the aid of MD simulations, we show (1) Born's stability criteria are valid only in the case of zero external stress, and (2) his thermoelastic melting criterion, with some modification, is valid for the homogeneous process (mechanical melting or upper limit of superheating) that can occur when the free-energy–based heterogeneous process (melting by nucleation and growth) is kinetically suppressed. These and related results on crack nucleation, pressure-induced polymorphic transition, and amorphization point to the fundamental role of elastic instabilities in triggering unstable structural responses of homogeneous crystals.

Introduction

In 1939, Born set forth a simple criterion for crystal melting, namely, melting should be accompanied by the loss of shear rigidity (Born 1939). Expressed in terms of the shear modulus G for a cubic crystal, the melting point T_m is that temperature at which

$$G(T_m) = 0. \tag{4.1}$$

A year later, he extended this concept to lattice deformation (Born 1940) by deriving the well-known conditions for structural stability for cubic crystals

$$C_{11} + 2C_{12} > 0, \quad C_{11} - C_{12} > 0, \quad C_{44} > 0 \tag{4.2}$$

where C_{11}, C_{12}, and C_{44} (= G) are the elastic constants (Voigt notation).

Our purpose is to examine on what basis Born's criteria for melting and lattice stability may be considered valid. Shortly after equation (4.1) was proposed, experimental results obtained on NaCl single crystals showed the two shear constants, C_{44} and $(C_{11} - C_{12})$, have nonzero values at the melting point (Hunter and Siegel 1942). Moreover, it was not clear how equation (4.1) could explain the presence of latent heat and volume change associated with a first-order thermodynamic phase transition. The status regarding criteria equation (4.2) seems to be more ambiguous; neither stringent tests have been performed nor qualifications discussed concerning its possible limitations.

One can appreciate the considerable challenge of ascertaining whether such criteria are capable of predicting the actual onset of an instability. The difficulty, on the theoretical side, was stability analyses have been formulated in different ways (Hill 1975; Hill and Milstein 1977), and few explicit calculations of elastic constants at the critical condition have been reported to make an unambiguous test. On the experimental side, competing effects frequently render the determination of the triggering instability uncertain. It is fair to say while the shortcomings of equation (4.1) are well known, the predictive value of equation (4.2) has mostly gone unscrutinized.

The approach that we consider begins with the results of a corresponding set of elastic stability criteria for homogeneous lattices under arbitrary but uniform external load that we have derived (Wang et al. 1993; Wang et al. 1995). Combining the new criteria, which may be considered as generalizations of equation (4.2), with direct MD simulation of structural response of a perfect crystal to pure dilatation, we can show that equation (4.2) is not valid except for vanishing external stress. Instead of testing equation (4.1) using experimental data, we propose to analyze and simulate the process of isobaric heating of a perfect crystal without surfaces or defects of any kind. For this case, equation (4.2) would be applicable if heating were carried out at zero pressure. As we see below, MD simulation shows that at the onset of melting, one of the shear constants indeed vanishes, although it is $(C_{11} - C_{12})$ rather than C_{44}. Moreover, the observed melting temperature, or equivalently the critical lattice strain, is in remarkable agreement with predictions based on the new criteria. Since the system we analyze and simulate is a defect-free lattice with no free surfaces, the melting that is being observed here does not pertain to the free-energy–based heterogeneous process involving nucleation and growth. That process, if not kinetically suppressed in simulation by virtue of elimination of all defects and surfaces, would set in at a lower temperature (the physical or actual melting point) and thereby preclude the melting process associated with an elastic instability. Allowing for these modifications, we conclude that Born's concept of thermoelastic mechanism of melting, which should be applied only to mechanical and not thermodynamic melting, has been reconciled with his lattice stability criteria.

Elastic Stability Criteria at Finite Load

The criteria for elastic stability of a homogenous crystal under conditions of arbitrary external load have been derived by formulating a Gibbs integral, which combines the change in the Helmholtz free energy with the external work done during deformation (Wang et al. 1993; Wang et al. 1995). One then sees clearly Born's derivation amounts to neglecting the contribution from the external work. It therefore follows that predictions of critical load in general will depend on the nature of the applied stress, in contrast to criteria equation (4.2) involving only the elastic constants. It is useful to regard the new criteria as generalizations of equation (4.2) in which the elastic constant tensor C is replaced by an elastic stiffness tensor B (Wallace 1972),

$$B = C + \Lambda \tag{4.3}$$

where the tensor Λ depends only on the applied stress σ. For cubic crystals and in the case of hydrostatic loading, $\sigma = -PI$, where I is the unit tensor and $P > 0$ for compression, the generalized stability criteria (Wang et al. 1993; Wang et al. 1995; Barrons and Klein 1965; Hoover, Holt, and Squire 1969) have the form

$$C_{11} + 2C_{12} + P > 0, \quad C_{11} - C_{12} - 2P > 0, \quad C_{44} - P > 0 \tag{4.4}$$

Even though equations (4.2) and (4.4) have very similar appearance, the presence of the external stress obviously can alter the competition between the different modes of instability to the extent the two equations can give different predictions, qualitatively and quantitatively.

A simple but conclusive demonstration of the validity of equation (4.4) is to consider a closed-packed lattice under pure dilatation ($P < 0$). Figure 4.1 shows the variation with lattice strain of the two moduli, $B_T = (C_{11} + 2C_{12})/3$ and $G' = (C_{11} - C_{12})/2$ in the case of Born's criteria in equation (4.2), and $B_T = (C_{11} + 2C_{12} + P)/3$ and $G' = (C_{11} - C_{12} - 2P)/2$ in the case of the generalized criteria of equation (4.4). The third modulus, that involving C_{44}, is not shown because it is not a competing mode of instability for this particular loading. These results are obtained by MD simulation using an interatomic potential model for the fcc metal gold (Foiles, Baskes, and Daw 1986) (details of the potential are of no interest here). The simulation cell is cubic and contains $N = 504$ atoms, arranged in an fcc structure with periodic border conditions. Starting with the lattice parameter, a, set at the equilibrium value a_0 (corresponding to zero pressure and minimum potential energy) and temperature $T = 500$ K, we equilibrate the system at incrementally larger values of a while maintaining constant temperature by velocity rescaling after every time step.

It can be seen in figure 4.1 that one would predict on the basis of equation (4.2) that the system would fail by the vanishing of the shear constant G' (the Born instability) at a critical strain around 1.025. On the other hand, according to equation (4.4), failure is predicted to be caused by the vanishing of the bulk modulus B_T (known as the spinodal instability) at a critical strain (extrapolated) of 1.059. To see which prediction is correct,

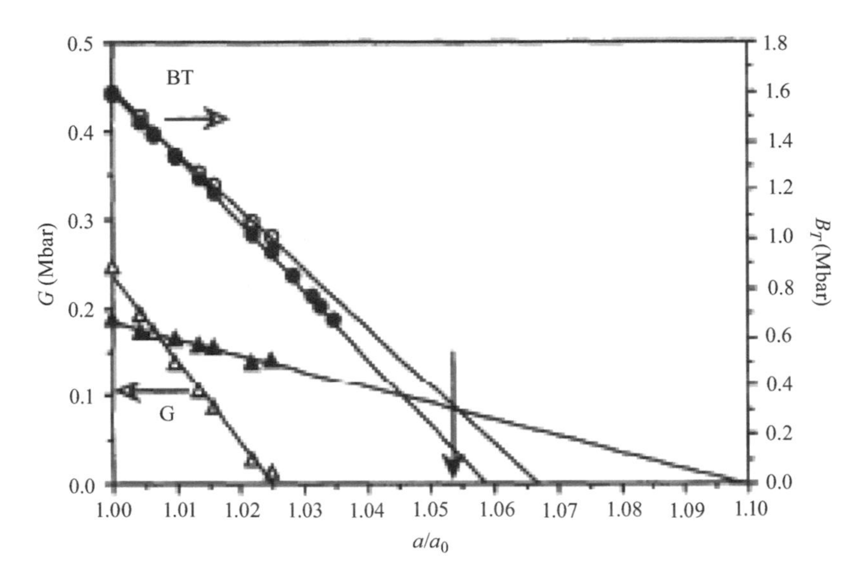

Figure 4.1
Variation of elastic moduli B_T and G' with lattice strain a/a_0 in an fcc lattice under pure dilatation, where a is the lattice parameter at temperature T and a_0 is the value at $T = 0$ K. Open and closed symbols denote results from equations (4.2) and (4.4), respectively. Solid lines are linear extrapolations to give critical values for each instability. The arrow indicates the critical strain observed by direct simulation. Figure from (Wang et al. 1997).

we have run the simulations up to the point of failure by dilatation. What was observed was that at a strain of 1.053, the system responded abruptly by a sudden release of internal stress with a corresponding lowering of the enthalpy, both being consequences of lattice decohesion in the form of cavitation. Thus, the validity of the generalized criteria in equation (4.4) is clearly established. It also follows from our finding that Born's criteria in equation (4.2) should not be used to determine the theoretical strength of a crystal (Born and Furth 1940; Macmillan and Kelly 1972).

Isobaric Heating to Melting at $P = 0$

Given that the generalized criteria of equation (4.4) obviously reduce to Born's results in the limit of zero load, then equation (4.2) is a valid description of lattice stability in the special case of a cubic crystal being heated to melting at zero pressure. This process is of interest because it can be studied by MD simulation in a straightforward manner and the results would provide a clean test of the melting hypothesis stated in equation (4.1).

For the simulation, we use the same model already described except that the simulation cell now contains $N = 1{,}372$ atoms and periodic border conditions are imposed in the manner of Parrinello and Rahman (Parrinello and Rahman 1981). A series of isobaric-isothermal simulations (with velocity rescaling) are carried out at various temperatures. At each temperature, the atomic trajectories generated are used to compute the elastic constants at the current state using appropriate fluctuation formulas (Ray 1988).

Figure 4.2 shows the variation with temperature of the lattice strain a/a_0 along the three cubic symmetry directions. The slight increase with increasing temperature merely indicates the lattice is expanding normally with temperature, and the results for the three directions are the same as they should be. At $T = 1{,}350$ K one sees a sharp bifurcation in the lattice dimension where the system elongates in two directions and contracts in the third. This is a clear sign of symmetry change, from cubic to tetragonal. To see whether the simulation results are in agreement with the prediction based on equation (4.2), we show in figure 4.3 the variation of the elastic moduli with temperature, or equivalently with lattice strain in view of the one-to-one correspondence shown in figure 4.2; the three moduli of interest are the bulk modulus $B_T = (C_{11} + 2C_{12})/3$, tetragonal shear modulus $G' = (C_{11} - C_{12})/2$, and rhombohedral shear modulus $G = C_{44}$. On the basis of figure 4.3, one would predict the incipient instability to be the vanishing of G', occurring at the theoretical or predicted lattice strain of $(a/a_0)_{th} = 1.025$. From the simulation at $T = 1{,}350$, the

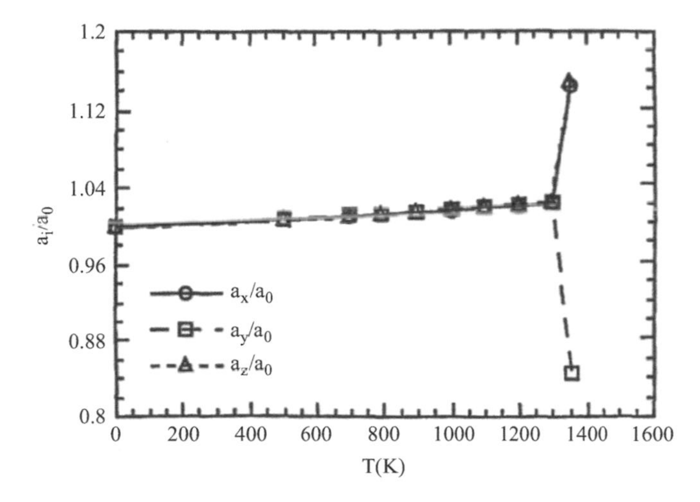

Figure 4.2
Variation of lattice strain a/a_0 with temperature along three Cartesian directions in the simulation of an isobaric ($P = 0$) heating process. Figure from (Wang et al. 1997).

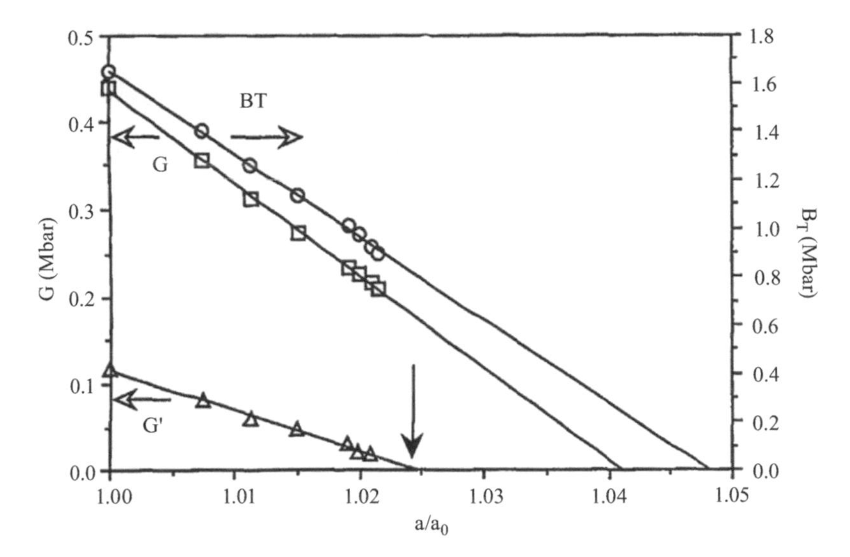

Figure 4.3
Variation of lattice strain a/a_0 with temperature along three Cartesian directions in the simulation of an isobaric ($P = 0$) heating process. Figure from (Wang et al. 1997).

observed strain is $(a/a_0)_{obs} = 1.024$. Thus, we can conclude that the vanishing of tetragonal shear is responsible for the structural behavior.

For more details of the system behavior at $T = 1{,}350$ K, we show in figure 4.4 the time evolution of the lattice strain, the off-diagonal elements of the cell matrix H, and the system volume. It is clear from figure 4.4(a) that the onset of the $G' = 0$ instability triggers both a shear (figure 4.4b) and a lattice decohesion (figure 4.4c), the latter providing the characteristic volume expansion associated with melting. This sequence of behavior, which has not been previously recognized, implies that the signature of a first-order transition, namely, latent volume change, is not necessarily associated with the incipient instability. Our results also provide evidence supporting Born's picture of melting being driven by a thermoelastic instability (Born 1939). This is now being reinterpreted to involve a combination of loss of shear rigidity and vanishing of the compressibility (Boyer 1985). Moreover, it is essential to recognize that the thermoelastic mechanism can only be applied to the process of mechanical instability (homogeneous melting) of a crystal lattice without defects, and not to the coexistence of solid and liquid phases at a specific temperature (heterogeneous melting) (Lutsko et al. 1989; Wolf et al. 1990), as we will discuss in the companion essay (essay 5).

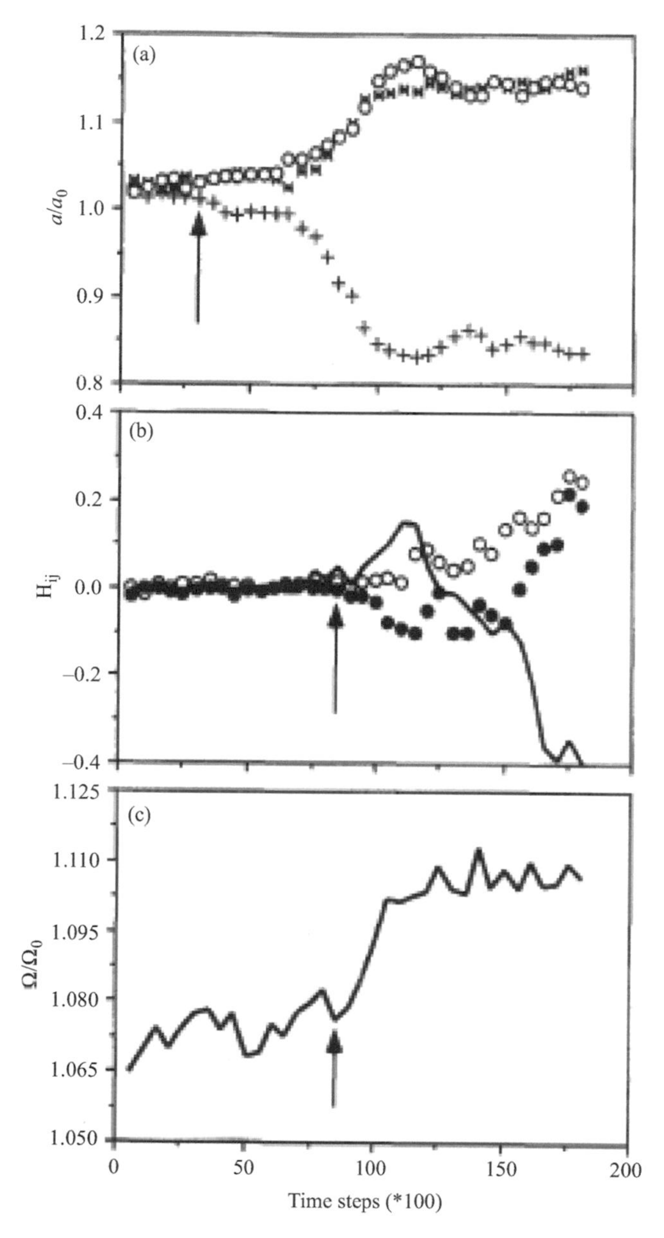

Figure 4.4
Time responses of (a) lattice strain along three initially cubic directions, (b) off-diagonal elements of the cell matrix H, H_{12}, H_{13}, H_{23}, (c) normalized system volume. Arrows indicate the onset of Born instability in (a), shear instability in (b), and lattice de-cohesion in (c). Figure from [Wang 1997].

Implications and Perspective

It is worthwhile restating what the combination of stability analysis and MD simulation has contributed to the understanding of Born's two criteria. That the stability criteria equation (4.2) are valid only under vanishing external load is quite clear, both theoretically and in simulation studies. Since it is often advantageous to be able to predict a priori the critical stress or strain for the onset of instability, the availability of equation (4.4) could facilitate more quantitative analysis of simulation results. While our results for an fcc lattice with metallic interactions show that homogeneous melting is triggered by $G' = 0$ and not equation (4.1), nevertheless, they constitute clear-cut evidence that a shear instability is responsible for initiating the transition. The fact that simulation reveals a sequence of responses all linked to the competing modes of instabilities (figure 4.4) implies it is no longer necessary to explain all the known characteristic features of melting on the basis of the vanishing of a single modulus. In other words, independent of whether $G' = 0$ is the initiating mechanism, the system will in any event undergo volume change and latent heat release in sufficiently rapid order (on the timescale of physical observation) that these processes are all associated with the melting phenomenon. Generalizing this observation further, one may entertain the notion of a hierarchy of interrelated stability catastrophes of different origins, elastic, thermodynamic, vibrational, and entropic (Tallon 1989).

Lastly it may be mentioned that in subsequent studies, the stability criteria in equation (4.4) have led to precise identifications of the elastic instability triggering a particular structural transition. In hydrostatic compression of Si, the instability which causes the transition from diamond cubic to β-tin structure is the vanishing of $G'(P) = (C_{11} - C_{12} - 2P)/2$ (Mizushima, Yip, and Kaxiras 1994). In contrast, compression of crystalline SiC in the zinc blende structure results in an amorphization transition associated with the vanishing of $G(P) = C_{44} - P$ (Tang and Yip 1995). For behavior under tension, crack nucleation in SiC (Tang and Yip 1994) and cavitation in a model binary intermetallic (Cleri, Wang, and Yip 1995), both triggered by the spinodal instability, vanishing of $B_T(P) = (C_{11} + 2C_{12} + P)/3$, are results that are analogous to the observations reported here. Also notice that in the present study a crossover from spinodal to shear instability can take place at sufficiently high temperature (Wang et al. 1995).

References

Barrons, T. H. K., and M. L. Klein. 1965. "Second-order elastic constants of a solid under stress." *Proceedings of the Physics Society* 85: 523.

Born, M. 1939. "Thermodynamics of crystals and melting." *Journal of Chemistry and Physics* 7: 591.

Born, M. 1940. "On the stability of crystal lattices. I." *Mathematical Proceedings of the Cambridge Philosophical Society* 36: 160.

Born, M., and R. Furth. 1940. "The stability of crystal lattices. III: an attempt to calculate the tensile strength of a cubic lattice by purely static considerations." *Mathematical Proceedings of the Cambridge Philosophical Society* 36: 454.

Boyer, L. L. 1985. "Theory of melting based on lattice instability." *Phase Transitions* 5: 1.

Cleri, F., J. Wang, and S. Yip. 1995. "Lattice instability analysis of a prototype intermetallic system under stress." *Journal of Applied Physics* 77: 1449.

Foiles, S. M., M. I. Baskes, and M. S. Daw. 1986. "Embedded-atom-method functions for the fcc metals Cu, Ag, Au, Ni, Pd, Pt, and their alloys." *Physical Review B* 33: 7983.

Hill, R. 1975. "On the elasticity and stability of perfect crystals at finite strain." *Mathematical Proceedings of the Cambridge Philosophical Society* 77: 225.

Hill, R., and F. Milstein. 1977. "Principles of stability analysis of ideal crystals." *Physical Review B* 15: 3087.

Hoover, W. G., A. C. Holt, and D. R. Squire. 1969. "Adiabatic elastic constants for argon. Theory and Monte Carlo calculations." *Physica* 44: 437.

Hunter, L., and S. Siegel. 1942. "The variation with temperature of the principal elastic moduli of NaCl near the melting point." *Physical Review* 61: 84.

Lutsko, J. F., D. Wolf, S. R. Phillpot, and S. Yip. 1989. "Molecular-dynamics study of lattice-defect-nucleated melting in metals using an embedded-atom-method potential." *Physical Review B* 40: 2841.

Macmillan, N. H., and A. Kelly. 1972. "The mechanical properties of perfect crystals I. The ideal strength." *Proceedings of the Royal Society of London A* 330: 291–309.

Mizushima, K., S. Yip, and E. Kaxiras. 1994. "Ideal crystal stability and pressure-induced phase transition in silicon." *Physical Review B* 50: 14952.

Parrinello, M., and A. Rahman. 1981. "Polymorphic transitions in single crystals: a new molecular dynamics method." *Journal of Applied Physics* 52: 7182.

Ray, J. R. 1988. "Elastic constants and statistical ensembles in molecular dynamics." *Computer Physics Reports* 8: 109.

Tallon, J. L. 1989. "A hierarchy of catastrophes as a succession of stability limits for the crystalline state." *Nature* (London) 342: 658.

Tang, M., and S. Yip. 1994. "Lattice instability in β-SiC and simulation of brittle fracture." *Journal of Applied Physics* 76: 2719.

Tang, M., and S. Yip. 1995. "Atomic size effects in pressure-induced amorphization of a binary covalent lattice." *Physical Review Letters* 75: 2738.

Wallace, D. C. 1972. *Thermodynamics of Crystals*. New York: Wiley.

Wang, J., S. Yip, S. R. Phillpot, and D. Wolf. 1993. "Crystal instabilities at finite strain." *Physical Review Letters* 71: 4182.

Wang, J., J. Li, S. Yip, S. Phillpot, and D. Wolf. 1995. "Mechanical instabilities of homogeneous crystals." *Physical Review B* 52: 12627.

Wang, J., J. Li, S. Yip, D. Wolf, and S. Phillpot. 1997. "Unifying two criteria of Born: elastic instability and melting of homogeneous crystals." *Physica A: Statistical Mechanics and its Applications* 240: 396.

Wolf, D., P. R. Okamoto, S. Yip, J. F. Lutsko, and M. Kluge. 1990. "Thermodynamic parallels between solid-state amorphization and melting." *Journal of Materials Research* 5: 286.

Further Reading

The primary reference on which this essay is based is J. Wang, J. Li, S. Yip, D. Wolf, S. Phillpot, "Unifying two criteria of Born: elastic instability and melting of homogeneous crystals," *Physica A: Statistical Mechanics and its Applications* 240, 396 (1997).

For an early study of thermodynamic melting in silicon and the effects of defects and surfaces, see S. R. Phillpot, S. Yip, and D. Wolf. "How do crystals melt?" *Computers in Physics* 3: 20 (1989).

Limoge, Y., A. Rahman, H. Hsieh, and S. Yip. 1988. "Computer simulation studies of radiation-induced amorphization." *Journal of Noncrystal Solids* 99: 75.

Deng, D., A. S. Argon, and S. Yip. 1989. "A molecular dynamics model of melting and glass transition in an idealized two-dimensional material—I." *Philosophical Transactions of the Royal Society* (London) A329: 549.

Deng, D., A. S. Argon, and S. Yip. 1989. "Topological features of structural relaxation in a two-dimensional model atomic glass—II." *Philosophical Transactions of the Royal Society* (London) A329: 575.

Deng, D., A. S. Argon, and S. Yip. 1989. "Kinetics of structural relaxations in a two-dimensional model atomic glass—III." *Philosophical Transactions of the Royal Society* (London) A329: 595.

Deng, D., A. S. Argon, and S. Yip. 1989. "Simulation of plastic deformation in a two-dimensional atomic glass by molecular dynamics—IV." *Philosophical Transactions of the Royal Society* (London) A329: 613.

Lutsko, J. F., D. Wolf, S. R. Phillpot, and S. Yip. 1989. "Molecular dynamics study of lattice-defect nucleated melting in metals using an embedded-atom-method potential." *Physical Review B* 40: 2841.

Wolf, D., P. R. Okamoto, S. Yip, J. F. Lutsko, and M. Kluge. 1990. "Thermodynamic parallels between solid-state amorphization and melting." *Journal of Materials Research* 5: 286.

Wang, J., S. Phillpot, D. Wolf, and S. Yip. 1993. "Intrinsic response of crystals to pure dilatation." *Journal of Alloys and Compounds* 194: 407.

de Koning, M., A. Antonelli, and S. Yip. 1999. "Optimized free energy evaluation using a single reversible scaling simulation." *Physical Review Letters* 83: 3973.

de Koning, M., A. Antonelli, and S. Yip. 2001. "Single-simulation determination of phase boundaries." *Journal of Chemistry and Physics* 115: 11025.

Yip, S., J. Li, M. Tang, and J. Wang. 2001. "Mechanistic aspects and atomic-level consequences of elastic instabilities in homogeneous crystals." *Materials Science and Engineering. A* 317: 236. Elastic stability criteria for a homogeneous lattice under arbitrary external loading combined with MD simulation allow a systematic analysis of competing structural transitions as in pressure-induced polymorphic and crystal-to-amorphous transitions and also in thermoelastic (homogeneous) melting.

Subramanian, S., and S. Yip. 2002. "Structural instability of uniaxially compressed alpha quartz." *Computer Materials Science* 23: 116.

5 Crystal-Melting Kinetics

MD simulation has proved to be a unique capability in computational materials research. We demonstrate its ability to probe the structural responses of a crystal undergoing melting transition, emphasizing the atomic-level details not available from experiments. In the kinetics of melting, one is able to distinguish homogeneous thermoelastic process of mechanical melting, discussed in the preceding essay, from the free-energy–driven process of thermodynamic melting, which involves nucleation and growth and is therefore *heterogeneous*. Physical insights are thereby extracted through direct control over crystal defects and analysis of simulation-unique data.

Introduction

Melting occurs at a temperature T_m when a crystalline substance undergoes a phase change from a solid to a liquid (melt). Despite its common occurrence, we still do not know the structural arrangements of the atoms or their characteristic motions prior to and during melting. Most theoretical studies, including molecular simulations of melting, generally do not consider the effects of extrinsic lattice imperfections, such as surfaces, dislocations, and grain boundaries (GBs). The role of lattice defects in the destruction of long-range order consequently remains to be clarified.

Our inability to see *how* melting occurs does not prevent us from knowing *why* it occurs and *when* this first-order phase transition should take place. According to thermodynamics, the melting point T_m is that temperature at which the solid and liquid phases coexist in equilibrium, the condition when the Gibbs free energies of the two phases are equal. It therefore follows that at temperatures above coexistence, the crystal is unstable. On the other hand, thermodynamics says nothing about the *mechanism* of melting, nor *the duration of* the process. These questions pertain to the *kinetics* of the phenomenon. For a complete understanding of melting, one needs to be concerned with both thermodynamics and kinetics.

As background to a discussion of melting, one should distinguish between intrinsic and extrinsic lattice defects. Intrinsic defects, such as lattice vacancies, are produced thermally. By contrast, extrinsic defects are usually thermodynamically metastable. Early ideas on melting have mostly considered only the effects of intrinsic defects. Alternatively, one could relate melting to a certain system behavior reaching a threshold value, for example, melting caused by the onset of an instability when the displacements during thermal vibration of the atoms exceed a certain threshold value (Lindemann 1910). Another postulate, emphasizing the elasticity of a crystal lattice, suggests melting arises from the onset of a mechanical instability, such as the vanishing of a shear modulus (Born 1939) (see essay 4). In other theories the spontaneous production of intrinsic lattice vacancies and arrays of dislocations near the melting point, was proposed to be responsible for the breakdown of long-range crystalline order (Cahn 1986). All of these considerations do not address explicitly the effects of surfaces, external or internal.

Experiments, on the other hand, showed melting generally proceeds from surfaces into the interior crystalline regions—a process requiring a finite amount of time. For example, it has been demonstrated that melting of silica, SiO_2, and phosphorous pentoxide, P_2O_5, are not homogeneous processes. Instead the liquid phase nucleates at free surfaces and GBs, from which it propagates into the crystal (Cormia, Mackenzie, and Turnbull 1963). Also, measurements have been reported that small single crystals of silver can be *superheated* above the melting temperature when coated with gold, which have very similar lattice structure but a significantly higher melting point (Daeges, Gleiter, and Perepezko 1986). These observations not only raise the question of whether melting can be regarded as a homogeneous process, but also point to the need for a direct observation of melting where surface effects can be explicitly isolated and properly attributed regarding their thermodynamic or kinetic origin.

Melting in Silicon

The choice of silicon (Si) for MD simulation of melting is motivated by three factors (Phillpot, Yip, and Wolf 1989). First, due to its covalent bonding, the crystal structure is cubic diamond, with four nearest neighbors. Upon melting, the tetrahedral coordination changes to an average six-fold coordination in the liquid. This change then can be used to distinguish between "crystalline" and "liquid" surroundings in the evolving atomic configurations during the onset of melting. Secondly, a good empirical interatomic potential is available for Si (Stillinger and Weber 1985). This is important for simulation realism. Lastly, extensive experimental information on melting and freezing of Si is known by virtue of the industrial significance of this elemental material.

In any study of melting, knowledge of the thermodynamic melting point, T_m, is essential. Although any classical interatomic potential function is at best an approximate description, it is most helpful to know a priori the interatomic potential being used give a realistic value of T_m. In this case, the Stillinger–Weber three-body potential gives $T_m = 1691 \pm 20$ K, compared to the experimental value of 1,683 K (Broughton and Li 1987).

Order Parameter for the Melting Transition

Before discussing the simulation results, an order parameter should be selected as a quantitative measure to characterize the crystalline and melt regions. For our purposes, the static structure factor $S(k)$, the Fourier transform of the distribution of bond-lengths (see essay 1), is an appropriate order parameter, so defined that its value lies between 0 and 1 (Phillpot, Yip, and Wolf 1989).

Figure 5.1 shows the evolution at 2,200 K of the atomic configuration of a Si bicrystal at four time instants. The color coding of the coordination of each atom is: perfect crystal surrounding $K = 4$ (red), liquid environment $K = 6$ (green), and a squeezed particle $K < 4$ (blue). The bicrystal has been pulled apart at the interface—GB (110) plane—vertically to demarcate the interfacial plane of atoms. It is seen that local disordering appears around the GB plane and a region of particles with liquid-like coordination begins to propagate into the bulk.

From results like figure 5.1 but at other temperatures, one can determine the speed at which the melt region is moving into the bulk. Figure 5.2 shows the speed of penetration of the melt region varies systematically with the temperature at which the bicrystal is being heated.

As seen in figure 5.2, the five data points extrapolate to a temperature of $1{,}710 \pm 30$ K at zero interface velocity. The fact that at this temperature the crystal and melt regions are at equilibrium is just the *thermodynamic condition for melting—the coexistence of solid and liquid phases*. Thus, we have demonstrated a simulation-based procedure for determining the thermodynamic melting point T_m. This procedure is effectively an alternative to direct free-energy evaluation, the latter giving a value of $1{,}691 \pm 20$ K (Broughton and Li 1987).

As a further step in the study figure 5.3 shows the evolution of a defect-free system (pristine crystal with no surfaces) at temperature 2,500 K. One readily observes the onset of melting across the entire sample. This is in direct contrast to figure 5.1, where melting initiated at the GB.

In investigating the high-temperature behavior of a bicrystal, the system was first equilibrated at 1,600 K. The temperature was then stepped up rather rapidly in intervals

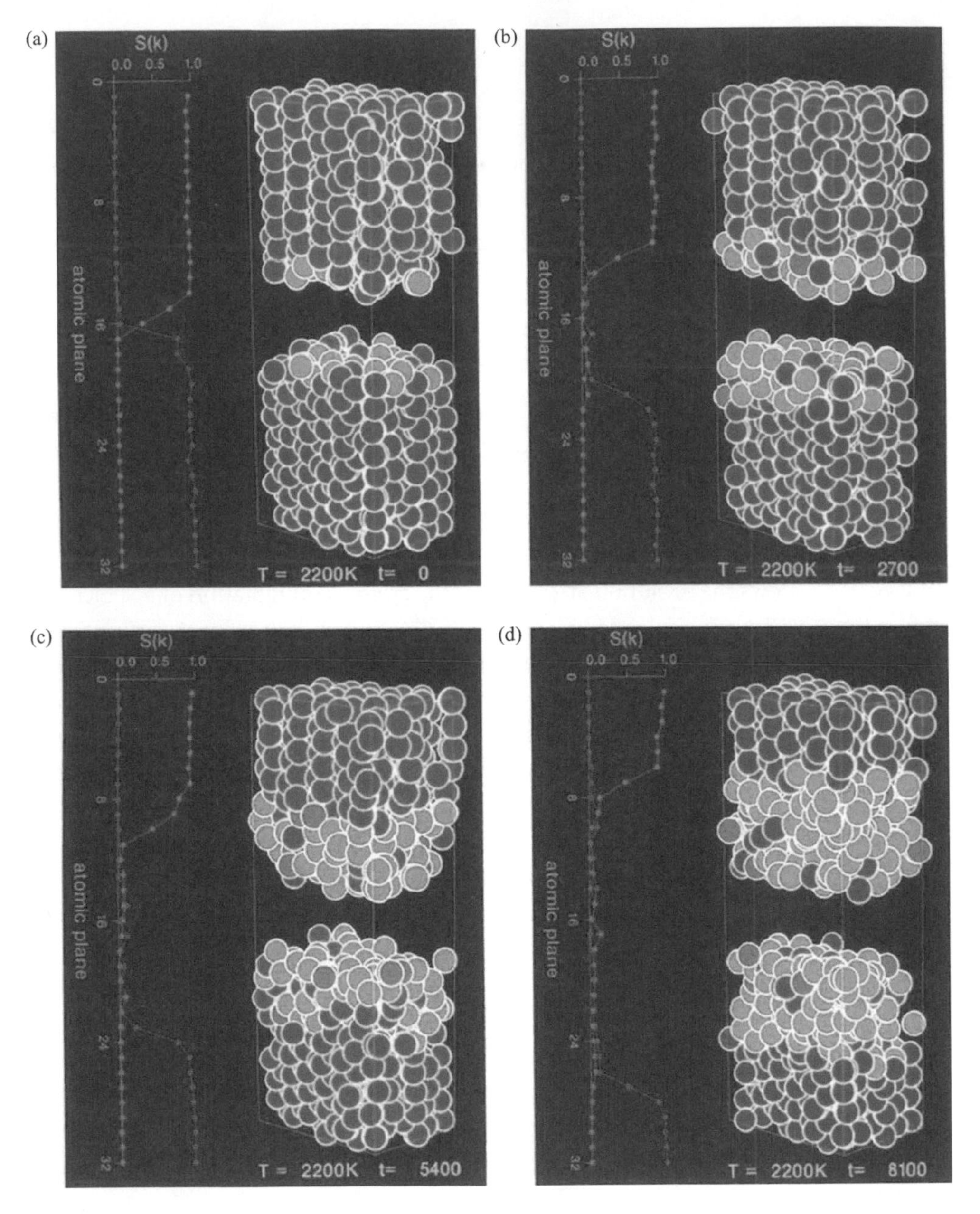

Figure 5.1

Time variation of coordination number distribution across a Si bicrystal at temperature 2,200 K and four equally spaced instants (a)–(d). The system initially is a pristine lattice except for a (110) GB located at the midsection. The upper and lower halves of the bicrystal have been pulled apart to display the GB interface more clearly. Particles with coordination number $K = 4$ (crystal environment), $K > 4$ (liquid-like surrounding), and $K < 4$ (more open surroundings) are coded red, green, and blue respectively. Reproduced from (Phillpot, Yip, and Wolf 1989), with the permission of AIP Publishing.

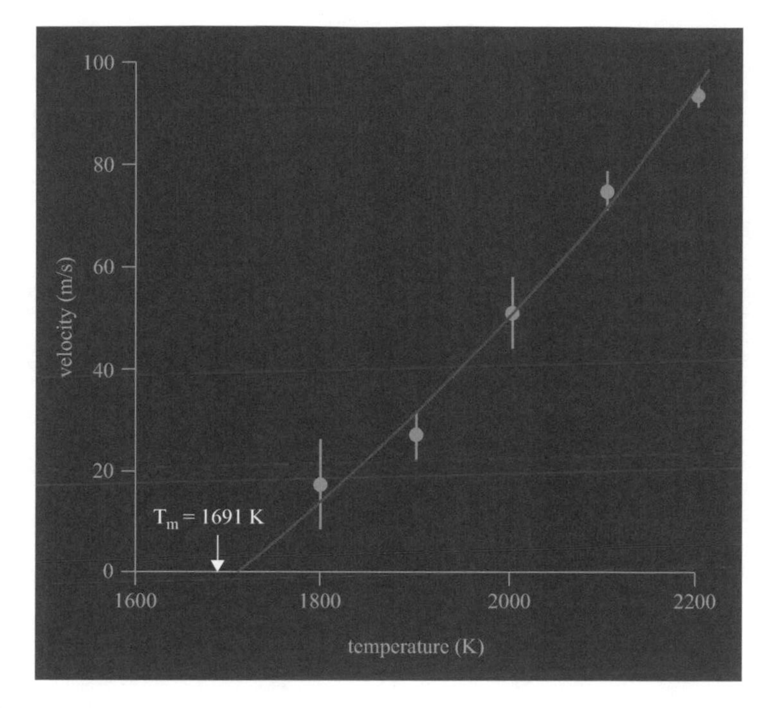

Figure 5.2
Variation of speed of crystal-melt interface with temperature at which the Si bicrystal has been heated. Extrapolation of the data using a cubic spline gives a temperature 1710 ± 30 K at which the speed vanishes (Phillpot, Yip, and Wolf 1989). The indicated temperature of 1,691 K is the predicted value of the thermodynamic melting calculated using the same interatomic potential as in the MD simulation (Broughton and Li 1987). Reproduced from (Phillpot, Yip, and Wolf 1989), with the permission of AIP Publishing.

of 100 K, allowing 100 time steps for approximate equilibration, until the desired final simulation temperature, ranging between 1,800 K and 2,200 K, was reached; this instant will be labeled $t = 0$. As already mentioned, the average atomic coordination increases from four to approximately six upon melting. To illustrate the different local environments of the atoms in the bulk, a surface or in the liquid, throughout our discussion red atoms will indicate perfect-crystal coordination (K = 4), with green atoms indicating K > 4 while for blue atoms K < 4.

We conclude from the above evidence that above T_m, the GB nucleates the liquid phase, which subsequently grows into the crystal, a process requiring thermally activated diffusion kinetics. The above simulations illustrate that every crystal, in principle, has two melting points: T_m and T_s. Conceptually the two transitions have distinct

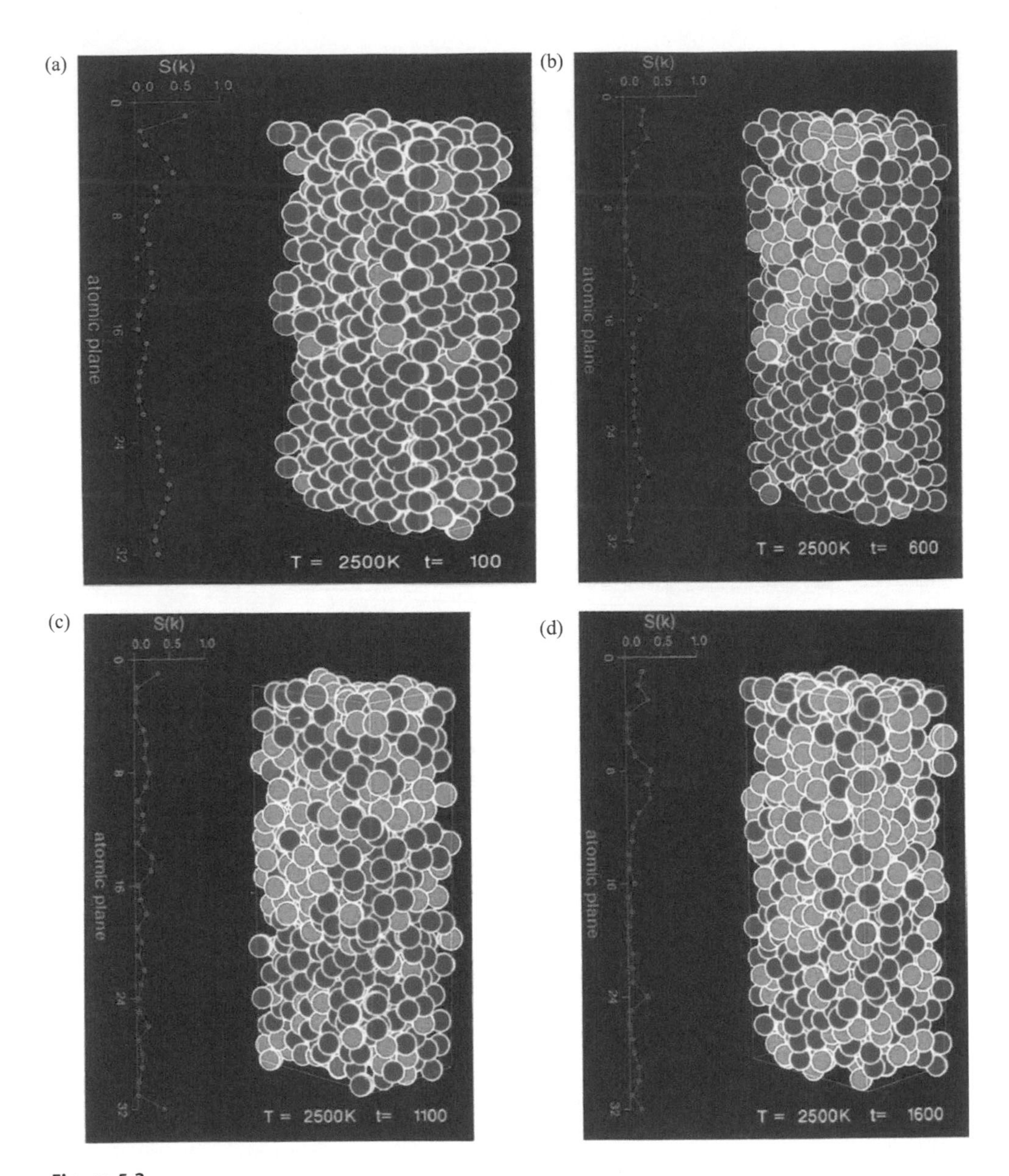

Figure 5.3
Heating of a pristine Si single-crystal sample with PBC (no free surface) showing spontaneous homogeneous melting across the entire sample. Same color coding of coordination as figure 5.1. Reproduced from (Phillpot, Yip, and Wolf 1989), with the permission of AIP Publishing.

physical origins: while *thermodynamic melting* is governed by the free energies of the liquid and the solid phases, *mechanical melting* is based on a phonon instability. Since at ambient pressure, the volume expansion required for mechanical melting is always larger than that associated with thermodynamic melting, the free energy always favors thermodynamic over mechanical melting; that is, $T_s > T_m$. However, as illustrated above, the former requires atomic mobility, and may therefore be kinetically hindered. If a crystal is melted under atmospheric conditions, the thermodynamic state variables usually will be such that high atom mobility in the liquid enables the nucleation and growth of the liquid phase at extended defects. However, if for example by uniformly expanding the crystal, melting is induced at a lower temperature, the consideration of limited atom mobility as a possible hindrance to phase change may be of significant importance. The crystal may, indeed, not be able to disorder at the volume specified by equilibrium thermodynamics until a larger volume is reached where the mechanical instability can occur.

There is considerable experimental evidence that solid-state amorphization, the process in which the long-range crystalline order is destroyed by external means (such as mechanical or chemical in nature, or by irradiation), can proceed by the same two distinct mechanisms as melting and that, in contrast to conventional melting, both types of transition can actually be observed. In a typical melting experiment, the order-disorder transition is induced by increasing the temperature (T) under ambient pressure (P), thus allowing the volume (V) to expand, a procedure guaranteeing high atom mobility at the point (T, $P[V]$) in thermodynamic phase space where the transition can occur. In a typical solid-state amorphization experiment, by contrast, the temperature is held fixed at some relatively low value, well below T_m. The role of the irradiation, or of the mechanical or chemical means, in inducing the crystal-to-amorphous transition is to expand the crystal lattice to the coexistence point in phase space where the thermodynamic transition can, in principle, occur. However, relatively low atom mobility gives rise to a competition between the heterogeneous and homogeneous processes, a competition governed by the level of atomic mobility at that point in phase space. Hence, while at higher temperatures mechanical amorphization will be preempted by the thermodynamic type of transition, at lower temperatures, this type of transition may be kinetically hindered due to the reduced atom mobility. However, at an even larger volume expansion than that at the thermodynamic coexistence point at a fixed temperature, the ultimate stability limit of the crystal may be reached, thus enabling a fast, homogeneous transition into the liquid state. Due to the low atomic mobility this noncrystalline state appears to be solid, although it merely represents a kinetically arrested liquid.

Discussion

The essential findings of this and the preceding essay may be summarized by three distinguishing characteristics of thermodynamic and mechanical melting:

1. While thermodynamic melting is based on the *free energies* of the crystalline and liquid states, mechanical melting is triggered by a *phonon instability*.
2. Thermodynamic melting is a *heterogeneous* process involving nucleation and growth of the liquid phase at extended defects. In contrast mechanical melting takes place *homogeneously*, without the need for the presence of lattice defects.
3. The growth of the liquid phase into the crystal (propagation of solid-liquid interfaces) requires thermally activated *diffusion kinetics* in the liquid. Mechanical melting, on the other hand, takes place typically within a few lattice vibration periods independent of temperature.

The present study illustrates the unique features of atomistic simulation discussed in essay 2, namely the abilities to precisely prescribe the initial system configuration, control the dynamic environment during simulation, and follow the system response in complete detail, all at the atomic level. In future simulation studies of complex materials phenomena, these capabilities will be further exploited. However, one should keep in mind that the fidelity of molecular simulation will always limited by the accuracy of the interatomic potential function used. For this reason, one should look for physical *insights* rather than the numerical results.

References

Born, M. 1939. "On the stability of crystal lattices. I." *Journal of Chemistry and Physics* 7: 591.

Broughton, J. Q., and X. P. Li. 1987. "Phase diagram of silicon by molecular dynamics." *Physical Review* B35: 9120.

Cahn, R. W. 1986. "Materials science: melting and the surface." *Nature* 323: 668.

Cormia, R. L., J. D. Mackenzie, and D. Turnbull. 1963. "Kinetics of melting and crystallization of phosphorus pentoxide." *Journal of Applied Physics* 34: 2239.

Daeges, J., H. Gleiter, and J. H. Perepezko. 1986. "Superheating of metal crystals." *Physics Letters A* 119: 79.

Lindemann, F. 1910. "The calculation of molecular Eigen-frequencies." *Zeitschrift für Physiks* 11: 609.

Phillpot, S., S. Yip, and D. Wolf. 1989. "How do crystals melt?" *Computers in Physics* 3: 20.

Stillinger, F. H., and T. A. Weber. 1985. "Computer simulation of local order in condensed phases of silicon." *Physical Review B* 31: 5262.

Further Reading

Phillpot, S. R., J. Lutsko, S. Yip, and D. Wolf. 1989. "Molecular dynamics study of lattice-defect nucleated melting in silicon." *Physical Review* 40: 2831.

Nguyen, T., P. S. Ho, C. Nitta, and S. Yip. 1992. "Thermal structural disorder and melting at a crystalline interface." *Physical Review B* 46: 6050. An extension to GBs showing no premelting although local disordering does take place at $T < T_m$.

Justo, J. F., M. Z. Bazant, E. Kaxiras, V. V. Bulatov, and S. Yip. 1998. "Interatomic potential for silicon defects and disordered phases." *Physical Review B* 58: 2539.

de Koning, M., A. Antonelli, and S. Yip. 1999. "Optimized free energy evaluation using a single reversible scaling simulation." *Physical Review Letters* 83: 3973.

Jin, Z. H., P. Gumbsch, K. Lu, and E. Ma. 2001. *Physical Review Letters* 87: 055703. A later study putting the melting phenomenon in a materials deformation context.

III Strength, Deformation, Toughness

6 Atomistic Measures: Strength and Deformation

Multiscale materials modeling is a foundational concept in computational materials research. Atomistic simulation plays a unique role in connecting the four characteristic scales from electronic structure to continuum mechanics. It defines the microscale where fundamental understanding resides along with microstructure manipulation—see figure P.1 in the prologue. Applications at this level have focused on stability and deformation of solids under mechanical loading, as well as mechanism insights that cannot be derived from experiments. Theoretical strength is defined through elastic modes of instability, or more generally, through the onset of soft vibrational modes in the deformed lattice. MD simulation of stress-strain response provides a direct measure of microstructure-sensitive strength of crystalline, amorphous, and nanocrystalline phases. Dislocation processes are systematically probed through characteristic mechanisms of nucleation and mobility as they occur in metals and semiconductors, including the distinction between slips versus twinning. From these case studies, one gains a broad appreciation of the capabilities of atomistic simulation in computational materials research.

Introduction

The innovation of materials in the global enterprise of science and technology has never been more relevant to global societal well-being. Advances in the synthesis of nanostructures and in high-resolution microscopy are enabling the discovery and detailed analysis of assemblies of atoms and molecules at a level unimagined only a short time ago. Another factor is the advent of large-scale computation, once a rare and sophisticated resource accessible only to a privileged few. In the current environment, multiscale materials modeling is a mature multidisciplinary research enterprise encompassing a wide range of physical structures and phenomena.

The foundation of computational materials, as introduced in the prologue, lies in the concept of multiscale modeling and simulation where conceptual models and simulation techniques are linked across the micro-to-macro length and timescales.

This essay is concerned with case studies of deformation behavior, defect nucleation, and mobility, all to demonstrate the capabilities of atomistic simulation for probing molecular mechanism. Through "unit process" problems, one can build the base for a bottom-up approach to understanding mechanical properties and performance. If the objective is to improve materials already in use, it would be sensible to start with a specific functional behavior in question. The challenge is then similar in spirit to solving an inverse problem.

Elastic Limits to Strength: Homogeneous Deformation

Understanding materials strength at the atomic level has been a goal in practically all disciplines of science and engineering. The theoretical basis for describing the mechanical stability of a crystal lattice lies in the formulation of stability conditions which specify the critical level of external stress that the system can withstand. Lattice stability is not only one of the most central issues in elasticity, it is also fundamental in any analysis of structural transitions in solids, such as polymorphism, amorphization, fracture, or melting.

Stability Criteria

Recall from essay 4, M. Born first showed that expanding the internal energy of a crystal in a power series in the strain and requiring positivity leads to conditions on the elastic constants if structural stability of the lattice is to be maintained (Born 1940; Born and Huang 1956). This concept of ideal strength of perfect crystals has been examined by Hill (Hill 1975) and Hill and Milstein (Hill and Milstein 1977), as well as used in various applications (Kelly and Macmillan 1986). That Born's results are valid only when the solid is under zero external stress has been explicitly pointed out in a later derivation by Wang et al. (Wang et al. 1995) invoking the formulation of a Gibbs integral. Further discussions were given by Zhou and Joos (Zhou and Joos 1996) and by Morris and Krenn (Morris and Krenn 2000), the latter emphasizing the equivalence to a thermodynamic formulation given by Gibbs' original formulation (Langely 1928). A consequence of these investigations is the clarification that theoretical strength can vary with the symmetry and magnitude of the applied load, rather than being an intrinsic property of the material system only. In this respect, the study of theoretical limits to material strength using atomistic models, including first-principles calculations (Morris et al. 2000), promises to yield new insights into mechanisms of structural instability.

Consider a perfect lattice undergoing homogeneous deformation under an applied stress τ, where the system configuration changes from X to $Y = JX$, with J being the deformation gradient or the Jacobian matrix. The associated Lagrangian strain tensor is

$$\eta = (1/2)(J^T J - 1) \tag{6.1}$$

Let the change in the Helmholtz free energy be expressed by an expansion in η to second order

$$\Delta F = F(X, \eta) - F(X, 0) = V(X)[t(X)\eta + (1/2)C(X)\eta\eta] \tag{6.2}$$

where V is the volume, t is the conjugate stress also known as the thermodynamic tension or the second (symmetric) Piola–Kirchhoff stress, and C is the fourth-order elastic constant tensor. For the work done by an applied stress τ, commonly called the Cauchy or true stress, we imagine a virtual move near Y along a path where $J \rightarrow J + \delta J\ J \rightarrow J + \delta J$, which results in an incremental work

$$\begin{aligned} \delta W &= \oint_S \tau_{ij} n_j \, \delta u_i \, dS \\ &= V(Y)\frac{\tau_{ij}}{2}\left(\frac{\partial u_i}{\partial Y_j} + \frac{\partial u_j}{\partial Y_i}\right) \\ &= V(Y)Tr(J^{-1}\tau J^{-T}\delta\eta) \end{aligned} \tag{6.3}$$

The work done over a deformation path ℓ, $\Delta W(\ell)$, is the integral given by equation (6.3) over the path. To examine the lattice stability at configuration X, we consider the difference between the increase in Helmholtz free energy and the work done by the external stress

$$\begin{aligned} \Delta G(Y,\ell) &= \Delta F(X,\eta) - \Delta W(\ell) \\ &= \int_\ell g(Y)\,d\eta \end{aligned} \tag{6.4}$$

where

$$g(Y) = \frac{\partial F}{\partial \eta} - V(Y)J^{-1}\tau J^{-T} \tag{6.5}$$

One may also interpret ΔG in the spirit of a virtual work argument. If the work done by the applied stress exceeds that which is absorbed as the free energy increase, then an excess amount of energy would be available to cause the displacement to increase and the lattice would become unstable.

We regard ΔG as a Gibbs integral in analogy with the Gibbs free energy. Notice that in general, ΔG depends on the deformation path through the external work contribution. This means that strictly speaking it is not a true thermodynamic potential on which one can perform the usual stability analysis. Nevertheless, $-g(Y)$ can be treated as a force field in deformation space for the purpose of carrying out a stability analysis (Wang et al.

1995). Suppose the lattice, initially at equilibrium at X under stress τ, is perturbed to configuration Y with corresponding strain η. A first-order expansion of $g(Y)$ gives

$$g_{ij}(\eta) = V(Y)B_{ijkl}\eta_{kl} + \ldots \tag{6.6}$$

where, by using $V(Y) = V(X) \det|J|$, one obtains

$$\begin{aligned} B_{ijkl} &= C_{ijkl} - \left[\frac{\partial(\det|J|\, J_{im}^{-1}\tau_{mn}J_{nj}^{-1})}{\partial\eta_{kl}}\right]_{\eta=0,\,J=I} \\ &= C_{ijkl} + \Lambda_{ijkl}(\tau) \end{aligned} \tag{6.7}$$

with

$$\Lambda_{ijkl}(\tau) = (1/2)[\delta_{ik}\tau_{jl} + \delta_{jk}\tau_{il} + \delta_{il}\tau_{jk} + \delta_{jl}\tau_{ik} - 2\delta_{kl}\tau_{ij}] \tag{6.8}$$

δ_{ij} being the Kronecker delta symbol for indices i and j. The physical implication of equation (6.6) is that in deformation space the shape of the force field around the origin is described by B. The stability condition is then the requirement that all the eigenvalues of B be positive, or

$$\det|A| > 0 \tag{6.9}$$

where $A = (1/2)(B^T + B)$, with B being in general asymmetric (Wang et al. 1995). In cases where the deformation gradient J is constrained to be symmetric, as in certain atomistic simulations at constant stress, one can argue that the condition $\det|B| > 0$ is quite robust. Thus, lattice stability is governed by the fourth-rank tensor B, a quantity which has been called the elastic stiffness coefficient (Wallace 1972). It differs from the conventional elastic constant by the tensor $\Lambda\Lambda$, which is a linear function of the applied stress. The foregoing derivation clearly shows the effect of external work that was not taken into account in Born's treatment. In the limit of vanishing applied stress, one recovers the stability criteria given by Born (Born 1940).

In the present discussion, we consider only cubic lattices under hydrostatic loading in which case the stability criteria take on a particularly simple form

$$\begin{aligned} K &= (1/3)(C_{11} + 2C_{12} + P) > 0 \\ G' &= (1/2)(C_{11} - C_{12} - 2P) > 0 \\ G &= C_{44} - P > 0 \end{aligned} \tag{6.10}$$

where C_{ij} are the elastic constants at the current pressure P, $P > 0$ (< 0) for compression (tension). K is seen to be the isothermal bulk modulus, and G' and G the tetragonal and rhombohedral shear moduli respectively. The theoretical strength is that value of P for which one of the three conditions in equation (6.10) is first violated. A simple demonstration showing that the external load must appear in the stability criteria is to subject a crystal to hydrostatic tension by direct atomistic simulation using a reasonable interatomic potential (see essay 4). In this case, one finds the instability mode is the vanishing of K, whereas the Born criteria, equation (6.10) with P set equal to zero, would predict the vanishing of G' (Wang et al. 1993).

It is worth mentioning that the six components of the eigenmodes of deformation corresponding to the three zero eigenvalues of det(B) are (1,1,1,0,0,0) $\delta\eta$, ($\delta\eta_{xx}$, $\delta\eta_{yy}$, $\delta\eta_{zz}$, **0, 0, 0**) with $\delta\eta_{xx}+\delta\eta_{yy}+\delta\eta_{zz}=\mathbf{0}$ in the order indicated in equation (6.10) (Wang et al. 1995). The deformation when the bulk modulus vanishes (spinodal instability) preserves the cubic symmetry, while for the tetragonal shear instability, the cubic symmetry must be broken.

The connection between stability criteria and theoretical strength is rather straightforward. For a given applied stress $\underline{\underline{\tau}}$, one can imagine evaluating the current elastic constants to obtain the stiffness coefficients B. Then by increasing the magnitude of $\underline{\underline{\tau}}$, one will reach a point where one of the eigenvalues of the matrix A (equation [6.9]) vanishes. This critical stress at which the system becomes structurally unstable is then a measure of theoretical strength of the solid. In view of this, one has a direct approach to strength determination through atomistic simulation of the structural instability under a prescribed loading. If the simulation is performed by MD, temperature effects can be taken into account naturally by following the particle trajectories at the temperature of interest.

Under a uniform load, the deformation of a single crystal is homogeneous up to the point of structural instability. For a cubic lattice under an applied hydrostatic stress, the load-dependent stability conditions are particularly simple, being of the form

$$B=(C_{11}+2C_{12}+P)/3>0, \quad G'=(C_{11}+C_{12}+2P)/2>0, \quad G=C_{44}-P>0 \tag{6.11}$$

where P is positive (negative) for compression (tension), and the elastic constants C_{ij} are to be evaluated at the current state. While this result is known for some time (Barrons and Klein 1965; Hoover, Holt, and Squire 1969; Basinski et al. 1970), direct verification against atomistic simulations showing that the criteria do accurately describe the critical value of P (P_c) at which the homogeneous lattice becomes unstable was not reported until later (Wang et al. 1993; Mizushima, Yip, and Kaxiras 1994; Tang and Yip 1994; Cleri 1995; Wang et al. 1995; Tang and Yip 1995). One may therefore regard P_c as a definition of theoretical or ideal tensile (compressive) strength of the lattice.

Turning now to MD simulations, we show in figure 6.1 the stress-strain response for a single crystal of argon under uniaxial tension at 35.9 K. At every step of fixed strain,

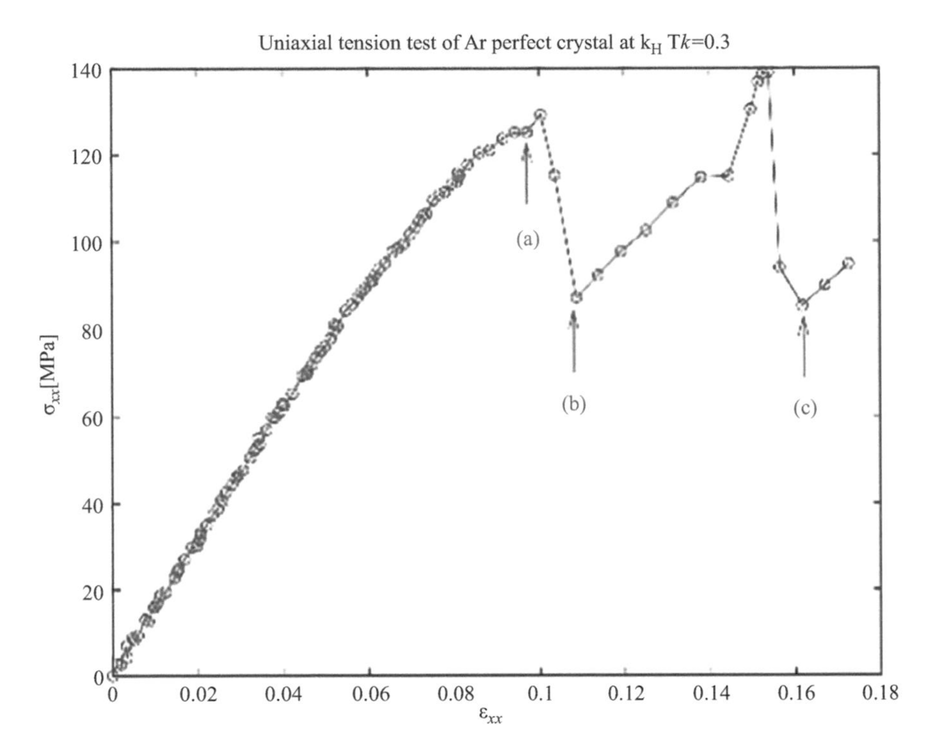

Figure 6.1
Atomistic stress-strain response of a single crystal of argon under uniaxial tensile deformation at constant strain at a reduced temperature of 0.3 (35.9 K). Simulated data are indicated as circles, and a dashed line is drawn to guide the eye. Reprinted from (Li et al. 2003) with permission from IOS Press. The publication is available at IOS Press through https://www.iospress.com/catalog/books/computational-materials-science

the system is relaxed and the virial stress evaluated. One sees the expected linear elastic response at small strain up to about 0.05; thereafter the response is nonlinear but still elastic up to a critical strain of 0.1 and corresponding stress of 130 MPa. Applying a small increment of strain beyond this point causes a dramatic stress reduction at point (b). Inspection of the atomic configurations at the indicated points shows the following.

At point (a), several point defects have been formed; one or more will act as subsequent nucleation sites for a larger defect eventually causing an abrupt stress relaxation. At the cusp, point (b), one can clearly discern an elementary slip on an entire (111) plane, the process being so sudden it was difficult to capture the intermediate configurations. We suspect a dislocation loop has been spontaneously created on the (111) plane that expanded at a high speed to join with other loops or inhomogeneities until it annihilates with itself on the opposite side of the periodic border of the simulation cell, leaving a stacking fault. As one increases the strain, the lattice loads up again until another slip occurs. At (c), one finds that a different slip system is activated. See essay 14 for a discussion of serrated flow in a metallic glass model.

Soft Modes

One may regard the stability criteria, equation (6.5), as manifestation in the long wavelength limit of the general condition for lattice vibrational stability. The vanishing of elastic constants then corresponds to the phenomenon of soft phonon modes in lattice dynamics. Indeed, one finds that under sufficient deformation, such soft modes do occur in a homogeneously strained lattice. To see the lattice dynamical manifestation of this condition, we apply MD to relax a single crystal sample with PBC at essentially zero temperature for a specified deformation at constant strain. The resulting atomic configurations are then used to construct and diagonalize the dynamical matrix. Figure 6.2 shows two sets of dispersion curves for the Lennard-Jones interatomic potential describing argon, which has an fcc structure, one for the crystal at equilibrium (for reference) and the other when the lattice is deformed under a uniaxial tensile strain of 0.138, which is close to the critical value (Li 2000). One can see in the latter a Γ′-point soft mode in the

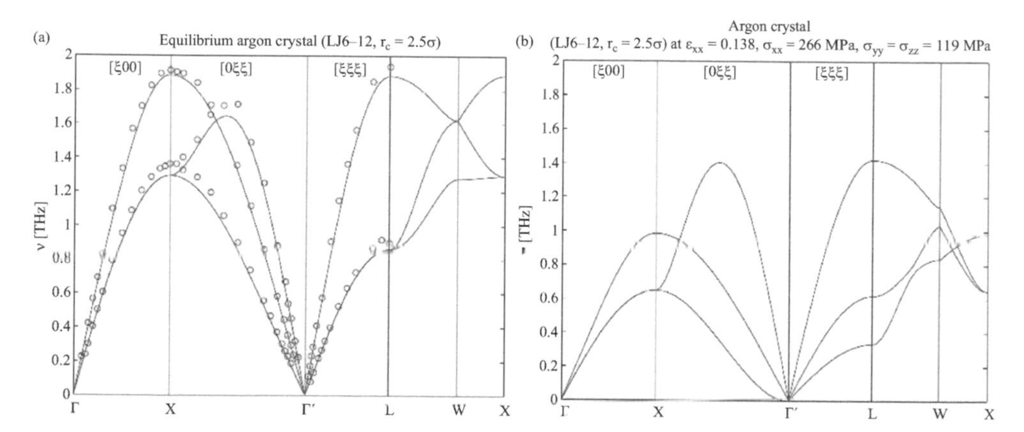

Figure 6.2
Phonon dispersion curves of single crystal of argon as described by the Lennard-Jones potential (lines): (a) comparison of results for equilibrium condition with experimental data (circles), (b) results for uniaxial tension deformation at strain of 0.138 (corresponding stresses of 266 MPa and 119 MPa along the tensile and transverse direction). Reprinted from (Li et al. 2003) with permission from IOS Press. The publication is available at IOS Press through https://www.iospress.com/catalog/books/computational-materials-science

(011) direction. Similar results for deformation under shear or hydrostatic tensile strain would show soft modes Γ′-point in the (111) direction and Γ-point in the (100) direction respectively. All these are acoustic zone-center modes; they would correspond to elastic instabilities. For a more complicated lattice such as SiC in the zinc blende structure, one would find that soft modes also can occur at the zone center (Li 2000). The overall implication is lattice vibrational analysis of a deformed crystal offers the most general measure of structural instability, and this again demonstrates that strength is not an intrinsic property of the material, rather it depends on the mode of deformation.

Defect Nucleation Mechanisms

Figure 6.1 is a typical stress-strain response on which one can conduct very detailed analysis of the deformation using the atomic configuration available from the simulation. This atomic-level version of structure-property correlation can be even more insightful than the conventional macroscopic counterpart simply because in simulation the microstructure can be as well characterized and visualized as one desires. As an illustration we repeat the deformation simulation using as initial structures other atomic configurations which have some distinctive microstructural features. Such studies have been reported on cubic SiC (3C or beta phase), which has a zinc blende structure, using an empirical bond-order potential (Tersoff 1989) and comparing the results for a single crystal and prepared amorphous and nanocrystalline structures (Li 2000). Figure 6.3 shows the stress-strain responses for hydrostatic tension at 300 K. At every step of fixed strain, the system is relaxed and the virial stress evaluated. Three samples are studied, all with PBCs, a single crystal (3C), an amorphous system that is an enlargement of a smaller configuration produced by electronic-structure calculations (Li and Yip 2002), and a model nanocrystal composed of four distinct grains with random orientations (7,810 atoms). As in figure 6.1, the single-crystal sample displays the expected linear elastic response at small strain up to about 0.03; thereafter, the response is nonlinear but still elastic up to a critical strain of 0.155 and corresponding stress of 38 GPa. Applying a small strain increment beyond this point causes a dramatic change with the internal stress suddenly relaxed by a factor of 4. Inspection of the atomic configurations (not shown) reveals the nucleation of an elliptical microcrack along the direction of maximum tension. With further strain increments, the specimen deforms via strain localization around the crack with essentially no change in the system-level stress.

The responses of the amorphous and nanocrystal SiC differ significantly from the single-crystal behavior just described. The former shows a broad peak, at about half

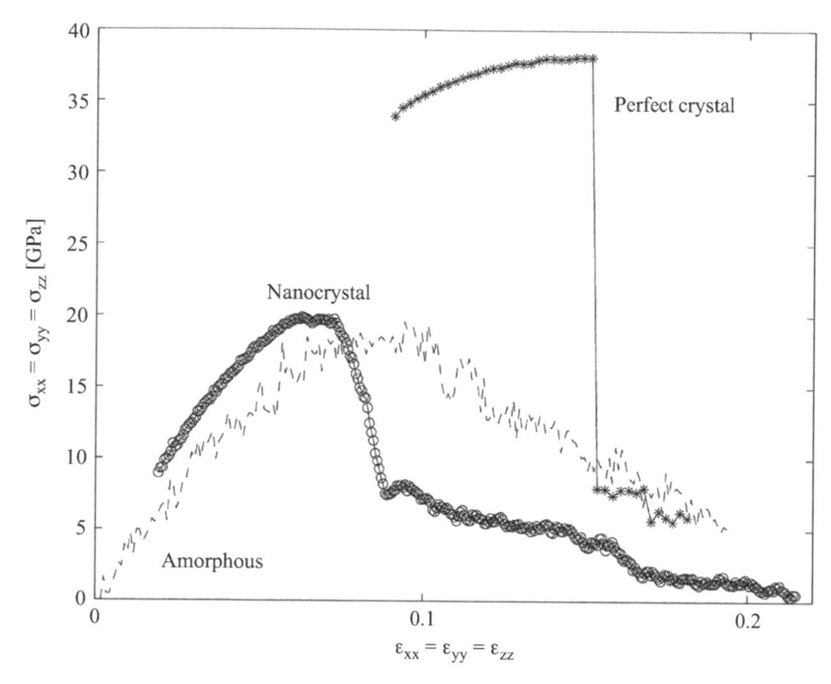

Figure 6.3
Variation of virial stress at constant strain from MD simulations of SiC (3C) under hydrostatic tension at 300 K in perfect crystal, amorphous, and nanocrystalline phases. Figure from (Li and Yip 2002).

the critical strain and stress, indicating a much more gradual structural transition. The deformed atomic configuration reveals channel-like decohesion at strain of 0.096 and stress 22 GPa. Another feature of the amorphous sample is that the response to other modes of deformation, uniaxial tension and shear, is much more isotropic relative to the single crystal, which is also understandable with bonding in SiC being strongly covalent and therefore directionally dependent. For the nanocrystal, the critical strain and stress are similar to the amorphous phase, except that the instability effect is more pronounced, qualitatively like that of the single crystal. The atomic configuration shows rather clearly the failure process to be intergranular decohesion. Overall, these observations allow us to correlate the qualitative behavior of the stress-strain responses with more detailed features of the system microstructure, namely, the local disorder (or free volume). This disorder is of course completely absent in the single crystal, while it is well distributed in the amorphous phase, and localized at the GBs in the nanocrystal. It can act as a nucleation site for structural instability, thereby causing a reduction of the critical stress and strain for failure (one might recall the effects of lattice defects on melting discussed in essay 5). Once a site is activated, it will tend to link up with neighboring activated sites, thus giving rise to different behavior between the amorphous and nanocrystal samples.

Incipient Plasticity: Nanoindentation

For *inhomogeneous* or localized deformation such as the phenomenon of nanoindentation, one expects local defects to be nucleated at certain sites in the system (the weak spots) when the system is driven across a saddle point on the potential-energy landscape (see essay 8). A continuum-level description of homogeneous nucleation was first explored by R. Hill (Hill 1962) through the concept of discontinuity of "acceleration waves." Later, J. R. Rice (Rice 1976) treated shear localization in much the same spirit and derived a formal criterion characterized by a tensor L playing the same role as the stiffness tensor B. This formalism can be taken to the discrete-particle level to obtain a spatially dependent nucleation criterion for practical implementation (Li et al. 2002; Zhu, Li, and Yip 2004a; Zhu, Li, and Yip 2004b). Consider a representative volume element (RVE) undergoing homogeneous deformation at finite strain to a current configuration x. Expanding the free energy F to second order in incremental displacement $u(x)$, one obtains

$$\Delta F = \frac{1}{2} \int_{V(\mathbf{x})} D_{ijkl} u_{i,j}(\mathbf{x}) u_{k,l}(\mathbf{x}) dV \tag{6.12}$$

where $D_{ijkl} = C_{ijkl} + \tau_{jl}\delta_{ik}$, τ_{ij} being the internal (Cauchy) stress, and $u_{i,j} \equiv \partial u_i(\mathbf{x})/\partial x_j$. In equation (6.12), C is the elastic constant tensor and $u(x)$ the strain at the current state of stress. By resolving the displacement as a plane wave, $u_i(\mathbf{x}) = w_i e^{i\mathbf{k}\cdot\mathbf{x}}$, one arrives at the stability condition for the RVE

$$\Lambda(\mathbf{w}, \mathbf{k}) = (C_{ijkl}\, w_i w_k + \tau_{jl}) k_j k_l > 0 \tag{6.13}$$

The structure of equation (6.13) is analogous to that of the stiffness tensor B (see equation [6.7]); the presence of the stress term represents the work done by the external load (Li et al. 2002). Whereas. B determines the overall crystal stability in homogeneous deformation, Λ is in contrast a site-dependent quantity, with its sign indicating the concavity of F. The significance of equation (6.13) is that if a pair of w, k exists such that Λ vanishes or becomes negative, then homogeneity of the RVE cannot be maintained and a defect singularity will form internally. In other words, the inequality can be used to interrogate the elastic stability of the RVE by minimizing Λ with respect to the polarization vector w and the wave vector k. The minimum value of Λ, $\Lambda_{\min}$, acts as a measure of the *local micro-stiffness*; wherever $\Lambda_{\min}$ vanishes, an instability is predicted at that spatial position.

Equation (6.13) is an energy-based criterion applicable to finite-strain deformation, with the minimization of Λ being the process where the local environment is sampled from point to point. Notice that the Helmholtz form of the free energy is used rather than the Gibbs form. This is because we are applying the plane wave resolution effectively in an RVE, the region containing the weak spot, with PBC, so no external work is involved (Li et al. 2004).

The usefulness of equation (6.13) has been demonstrated in analysis of MD simulation results on nanoindentation (Li et al. 2002a; Van Vliet et al. 2003; Zhu et al. 2004). The first few predictions of instability according to the Λ criterion were found to correspond remarkably closely with the jumps observed in the simulated load-displacement curve, as shown in figure 6.5. Because the simulations were performed at increments of fixed strain (displacement control mode), discontinuities appear as vertical jumps in contrast to figure 6.4.

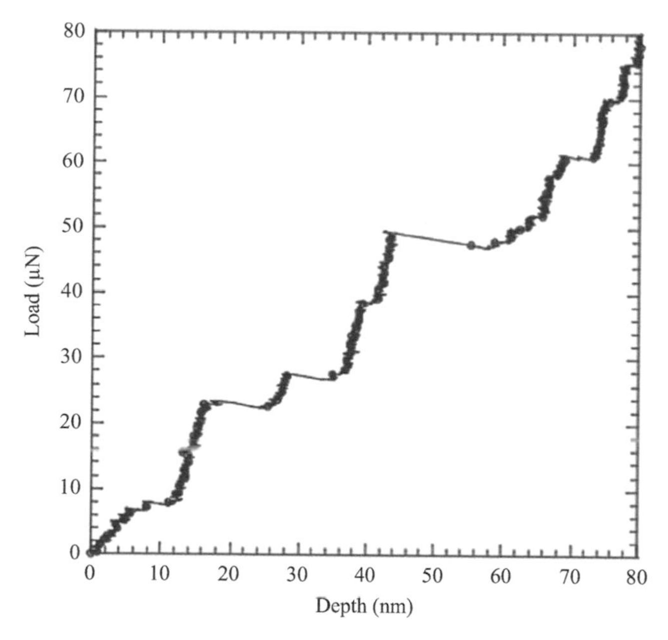

Figure 6.4
Typical load-displacement curve in a nanoindentation measurement (load control). Continuous increase of indentation depth with indenter loading shows elastic deformation interspersed with step-like horizontal jumps. These "bursts" indicate the onset of dislocation activity, each jump corresponding to a local instability in the system (not necessarily right under the indenter). Compare schematically with figure 14.1. Reprinted by permission from Springer Nature: (Yip 2006).

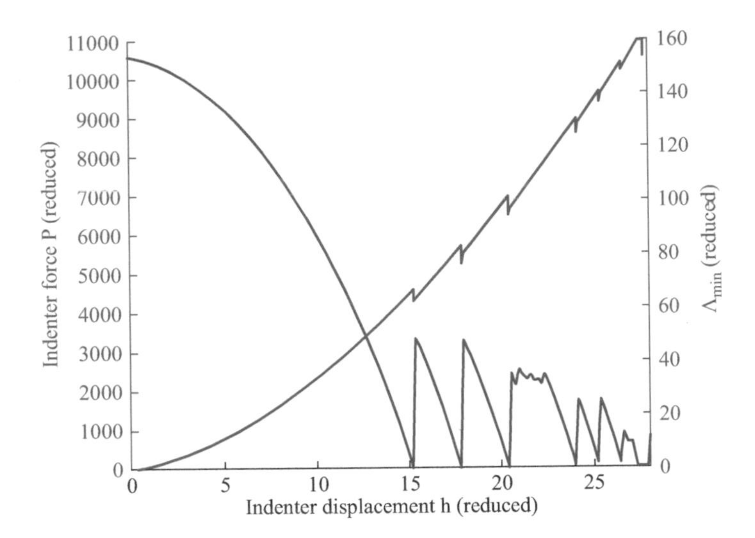

Figure 6.5
Load-displacement response in an MD simulation of nanoindentation (load control) on a slab of single crystal of copper (Cu) (rising curve), showing several vertical jumps similar to the dislocation bursts of figure 6.1. The decreasing curve gives the variation of Λ_{min} with indenter depth. According to equation (6.13), each vanishing of Λ_{min} signals a structural instability. Figure from (Li et al. 2002).

Instability in Shear Deformation: Dislocation Slip versus Twinning

Slip and twinning of crystal planes are competing processes by which a crystal can plastically accommodate large shear strains. They are distinguished by the number of planes which undergo sliding, as indicated schematically in figure 6.6. In the lattice response to shear, slip refers to *only one* relative displacement between two adjacent layers, while for twinning a stack of *three or more layers* must undergo uniform shear. When a crystal is being sheared uniformly, all the planes initially respond elastically. This would continue until symmetry is broken, when the system spontaneously transforms into a crystal containing a defect (loss of homogeneity). In practice considerations of factors such as the crystal structure and the material in question, the planes on which deformation is taking place, the temperature, and the shear rate are all considerations in predicting or expecting whether slip or twinning is favored. In the absence of such knowledge, one could resort to simulation for observing the atomistic details concerning system response near a saddle point.

MD simulations have been performed to observe the spontaneous nucleation of a deformation twin in a bcc crystal (Chang 2003; Chang et al. 2004). A simulation cell containing 500,000 atoms with PBCs is chosen to have the *X* (horizontal), *Y* (normal to plane of paper) and *Z* (vertical) axes oriented as shown in figure 6.7. Shear is applied

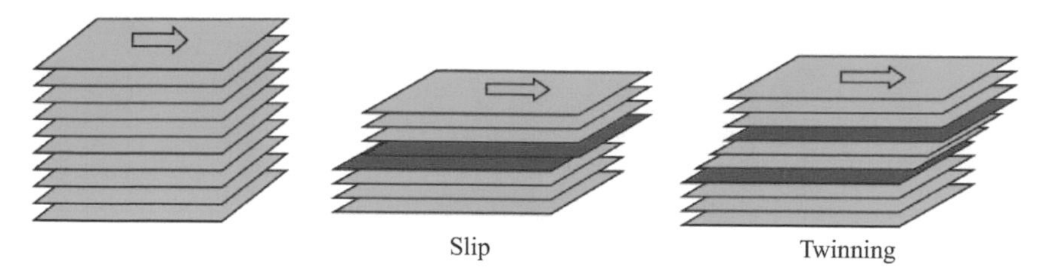

Figure 6.6
Schematic of a stack of undeformed crystal planes (left) about to undergo shear localization. A dislocation is nucleated by a *single* relative displacement (middle), whereas twinning requires *two or more* consecutive relative displacements (right, three displacements). Figure created by Sidney Yip.

at 3×10^6 s^{-1} on the *xy* plane in the *X* (twinning) direction. The strain rate is much too fast by laboratory standards but reasonable by MD standards. At 10 K, we observe homogeneous nucleation of a deformation twin at a critical shear stress of 12.2 GPa (7.8% strain). Once nucleation sets in, a sharp decrease in strain energy and shear stress is observed. The full three-dimensional (3D) configuration of the twinned crystal has too many degrees of freedom to be efficiently processed in any kind of detailed analysis. Given the twinned region is well localized, an overwhelming majority of the particles are not relevant for the characterization of defect configuration and energy. We therefore introduce a one-dimensional (1D) chain model to single out those coordinates essential to define the twin structure. As indicated in figure 6.7, we regard the twin as a 1D chain of "defect atoms" specified by a set of coordinates x_i, measured on the *X*-axis in the twinning direction. Each defect atom *i* represents a plane of physical atoms, the layer being perpendicular to the chain direction, with coordinates (x_i, y_i, z_i). In this model, the only degrees of freedom are the *relative displacements* in the twinning direction *between adjacent layers*, that is, $\Delta x_i = x_{i+1} - x_i$. For a twin consisting of *N* relative displacements of "defect atoms," there are *N primary* degrees of freedom, Δx_i, $i = 1, \ldots, N$, specifying the defect. All the other degrees of freedom, the relative displacements in the *Y* and *Z* directions will be considered frozen and not considered in the analysis. The system energy is therefore only a function of *N* variables, $E = E(0, \ldots 0, \Delta x_j, \ldots, \Delta x_{j+N}, 0, \ldots 0)$, with the displaced planes starting at position *j* and ending at $j + N$.

Using the chain model, with the energy functional $E(\Delta x_1, \ldots, \Delta x_N)$ evaluated numerically treating all the degrees of freedom except $\Delta x_1, \ldots, \Delta x_N$ in the 3D crystal as either frozen or under constrained relaxation, one has a means to examine the energetics of the twin defect for an arbitrary number of layers undergoing relative displacements. The simplest case is the one-component chain involving a rigid translation of the upper half of the lattice relative to the lower half. The 1D energy $E(0, \ldots, \Delta x_1, \ldots 0)$, allowing relaxation

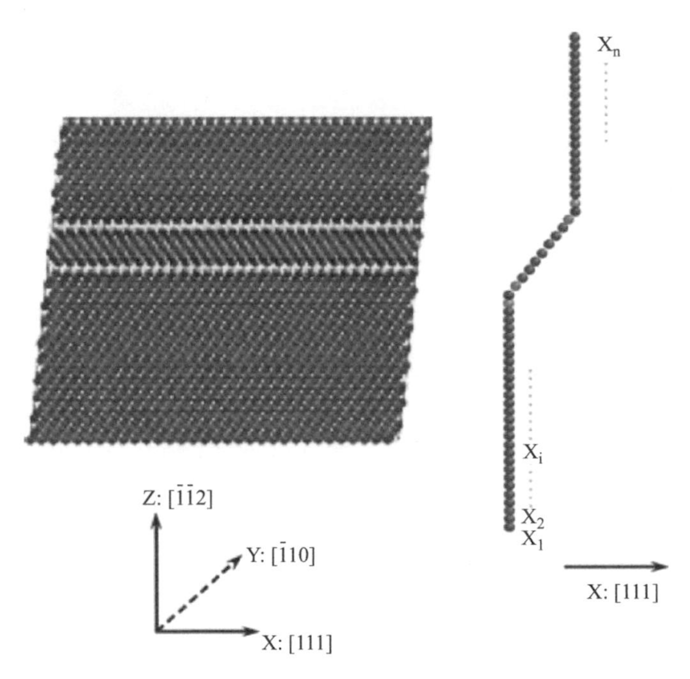

Figure 6.7
A 3D crystal in shear along the *X* axis containing a twinned region of seven planes, bounded by two twin boundaries (left). Schematic of the 1D chain model showing nine twinned layers (in this illustration) (right). Reprinted by permission from Springer Nature: (Yip 2006).

in the other two directions, is conventionally known as the γ-surface, a quantity commonly used to characterize lattice deformation in shear. The two-component chain is the one of present interest, as it is able to describe either a one-layer slip or a two-layer twin. The corresponding 2D energy surface is shown in figure 6.8, where a minimum is now seen around the displacements (b/3, b/3) (Chang et al. 2004). The significance is under positive shear the present system can either twin or slip. The energy barrier for twinning is found to be 0.672 eV with the saddle point at (0.36b, 0.16b), while for slip the barrier is 0.736 eV with the saddle point at (0.5b, 0.09b). Under negative shear, only slip is allowed, at a barrier of 0.808 eV. This is the kind of atomic-level energy landscape that enables one to analyze the competition between dislocation slip and deformation twin. Another useful way to examine the energy surface is in the form of a contour plot, also shown in figure 6.8. In this one can trace out the minimum-energy path for the two deformations, using any of the reaction pathway techniques in the literature. The particular result given here was obtained by the method of nudged elastic band (Jonsson, Mills, and Jacobsen 1998). The path is seen to connect the initial configuration at the perfect lattice energy

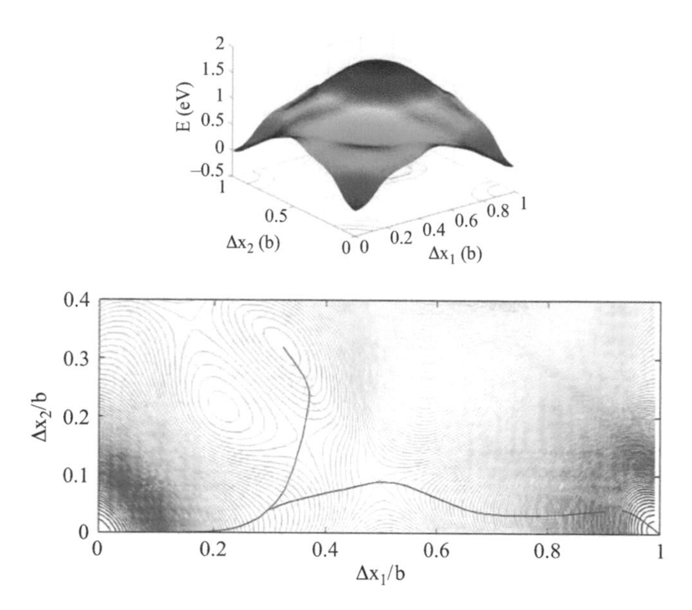

Figure 6.8
Strain energy surface (upper) and energy contour plot (lower) for the $(\bar{1}\bar{1}2)[111]$ deformation, with *Y* and *Z* relaxations, in a two-layer analysis. An energy minimum at relative displacements of *b*/3 for sliding between adjacent layers confirms the existence of a twin defect in the present model. Reprinted by permission from Springer Nature: (Chang et al. 2004).

minimum (0,0) with the two possible final-state configurations, an energy minimum corresponding to a two-layer twin at (*b*/3, *b*/3), and another minimum corresponding to a dislocation slip at (*b*, 0). The two paths bifurcate at (0.29*b*, 0.03*b*) before either of the saddle points is encountered. The system can either twin or slip after the bifurcation point; however, since the twinning path has a lower energy barrier than the slip path, 0.672 eV to 0.736 eV, twinning is expected to be favored.

Strain Localization in Twinning

In our MD simulation, the process that revealed the formation of a twin involved the application of a uniform shear strain to a perfect crystal in incremental steps. Initially, the system responded uniformly, the deformation being elastic and reversible; if the strain were released, the system would return precisely to its initial configuration. When the applied strain reaches a critical level, the system undergoes a structural transition whereby all the strain in the system is concentrated in the immediate surrounding of the defect, while the other parts of the crystal return to a state of zero strain. The initial and final configurations of such a transition are given in figure 6.9. To visualize

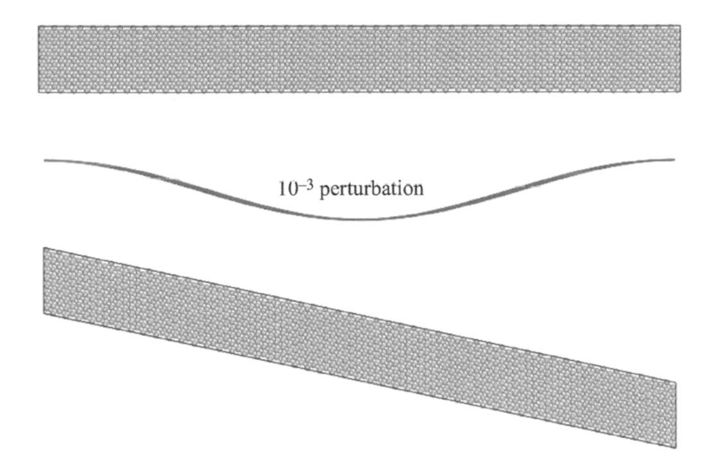

Figure 6.9
Nucleation of a deformation twin (shear localization), single crystal undeformed (upper) and in final sheared state (lower). Modulating perturbation wave (middle) follows the localization process by virtue of its distortion. Reprinted by permission from Springer Nature: (Yip 2006).

the rather complex sequence of atomic rearrangement that must take place, we introduced a simple tiling device in order to visualize the mechanism of the localization process (Chang 2003). The tiling procedure consists of imposing a modulating perturbation wave, a sinusoidal wave of small amplitude across the system, as shown by the thin curve in figure 6.9.

The perturbation wave should be so weak as to avoid any significant effect on the dynamical evolution of the system, while the wave distortion during localization helps to reveal the local atomic shuffling details that otherwise would be difficult to detect among all the particle displacements in the system. Metaphorically, the idea is like painting a line across a wall before it starts to collapse and then focusing only on the line as different parts of the wall start to crumble. By examining a video of the breakup of the perturbation wave, one is able to visually follow the sequence of structural transformations occurring in the nucleation of the deformation twin. In this case, what one sees can be described as a four-stage evolution scenario depicted in figure 6.10. In the initial stage, the linear wave grows in amplitude as the applied strain increases. When the amplitude reaches a certain level, nonlinearities set in and the wave starts to distort (steepen). As the distortion approaches the critical level, the wave front becomes increasingly sharp and then suddenly collapses into a shock front. In the final stage of strain localization, the wave settles into a profile indicating the presence of a twin.

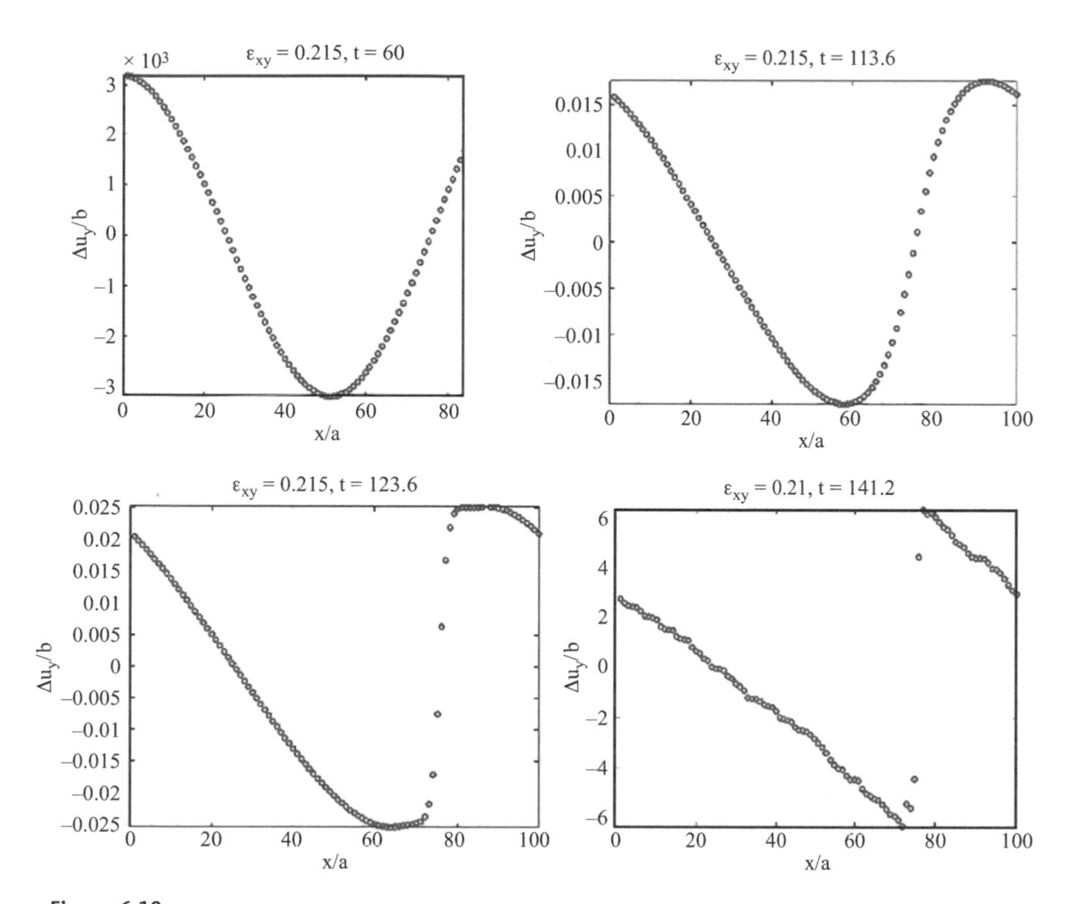

Figure 6.10
Evolution of the modulating perturbation wave during nucleation of a deformation twin. The wave distortions can be classified in four stages: growth of linear wave ($t = 60$), onset of nonlinearity ($t = 113.6$), increasing wave steepening toward a singularity ($t = 123.6$), and final state of strain localization ($t = 141.2$). Wave forms are those of the perturbation wave during the localization event. Reprinted by permission from Springer Nature: (Yip 2006).

One is reminded of a lattice deformation involving shear localization that is much more extensively investigated than nanoindentation and affine shear. This is the problem of crack tip behavior in a ductile solid. A recent simulation study using reaction pathway sampling has shown that a sharp crack in an fcc lattice, such as copper (Cu), will emit a dislocation loop under critical mode I (uniaxial tension) loading (Zhu, Li, and Yip 2004a). When the same method is applied to a brittle material, such as Si, a different result is found; the crack front advances by a series of bond breaking and reformation (Zhu, Li, and

Yip 2004b). Thus, the deformation response of a solid can be very sensitive to the nature of the chemical bonding. What happens when a relatively brittle material, a semiconductor or a ceramic, is subjected to nanoindentation or affine shear? Preliminary results indicate the system can undergo local disordering, suggesting yet another competing mechanism of local response to critical stress or strain. Clarification of this kind of phenomenon is work for the future. What is clear from the present discussion is *informed* atomistic simulations will continue to provide a wealth of structural, energetic, and dynamical information about the collective behavior of simple condensed matter, provoking in this way further advances in the statistical-physics and materials community.

Mechanisms of Defect Mobility

Single Dislocation Glide in a Metal

Most mechanical properties of crystalline materials are affected, to a greater or lesser extent, by the presence of dislocations. Dynamic properties of dislocations, in terms of intrinsic properties of the dislocation and the effects of temperature and the driving force, play important role in crystal plasticity (Hirth and Lothe 1982; Nadgornyi 1988). At zero temperature, continuous glide motion occurs only if the applied stress σ exceeds a critical value σ_c^o given by the maximum glide resistance. At finite temperatures macroscopic continuous glide can occur with the help of fluctuations at any stress up to a value $\sigma_c \leq \sigma_c^o$. In bcc metals, dislocation velocity can fall into four regions (Klahn, Mukherjee, and Dorn 1970; Neuhauser 1979). In the first region, $v/c_s \leq 10^{-5}$, c_s being the sound speed in the solid, the motion is controlled by the thermal release of dislocation from some equilibrium positions in the obstacle resistance profile. This is the thermally activated region where the motion is jerky and the stress dependence is highly nonlinear. Also, the velocity increases with temperature, and the motion is sensitive to imperfections. The second region, $10^{-5} < v/c_s < 10^{-2}$, is a continuation of the first, the motion being sensitive to all contributions to the glide resistance. In the third region where $v/c_s \geq 10^{-2}$, the drag resistance is predominant. Here $\sigma > \sigma_c$, the dependence of $v(\sigma)$ is linear, the velocity characteristically decreases with temperature and depends weakly on the imperfection concentrations. In the fourth region at $v \sim c_s$, relativistic effects can be observed.

To explore the enumerated regions through MD we consider a periodic simulation cell that contains an edge dislocation dipole, prepared as indicated in figure 6.11 (Chang, Bulatov, and Yip 1999). After first removing the atoms on two identical half planes, we displace the atoms immediately to the left and right of the gap by (111)/3 and allow the system to relax to zero stress. This gives a dislocation dipole configuration with Burgers vector $b = a/2(111)$ and glide plane (121). Simulation is carried out

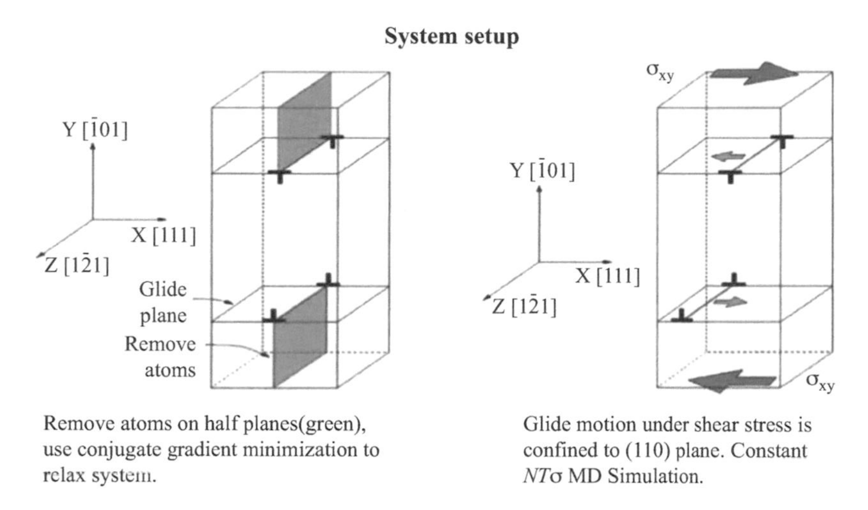

Figure 6.11
Periodic simulation cell containing an edge dislocation dipole under applied shear stress. The two dislocations will glide on the (121) slip plane. Reprinted by permission from Springer Nature: (Chang, Bulatov, and Yip 1999).

in the $NT\sigma$ ensemble, with an overall shear stress imposed on the simulation cell through the Parrinello–Rahman procedure (Parrinello and Rahman 1981).

To extract the dislocation velocity from the atomic trajectories, we need to locate the dislocation core position at each time step. Since the dislocation line is approximately parallel to the Z-axis, we divide the simulation cell into slices along this direction, with each slice containing one layer of atoms. For each slice, we first identify the two rows of atoms immediately above and below the slip plane. Then we calculate the *disregistry* between the two rows as a function of X-coordinate, keeping in mind that this function is periodic along X. The X-position of maximum *disregistry* is taken to be the dislocation core position. By averaging along the dislocation line, we determine the average line position at each time step, and from this information we deduce the dislocation velocity.

The dislocation velocity results obtained from a series of MD simulations on bcc metal molybdenum (Mo) at a temperature up to 150 K and stresses 5 MPa to 5 GPa are displayed in figures 6.12 and 6.13, showing the stress and temperature variations, respectively.

Figure 6.12 shows that in the stress range studied, the velocity varies essentially linearly with the stress, which is a distinctive feature of the third region discussed above. This behavior is almost to be expected in that MD is known to be restricted to microscopic distances and times; consequently, one can only probe the high-velocity,

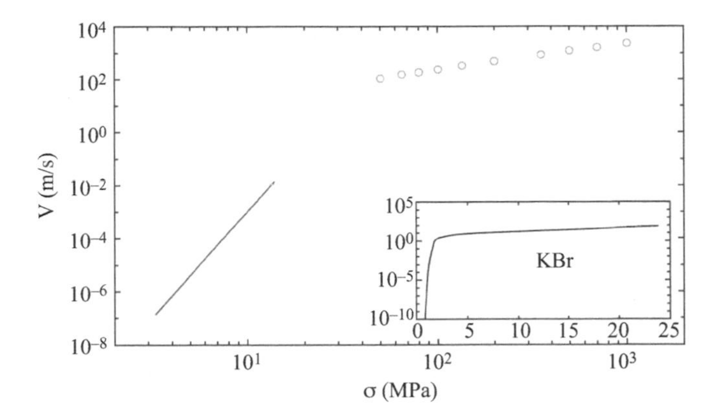

Figure 6.12
Stress dependence of edge dislocation velocity in Mo at $T = 77$ K. Simulation results shown as open circles and experimental results denoted by a solid line. Inset shows the change in dislocation velocities (edge and screw) as a function of shear stress in calcium-doped potassium bromide at room temperature where the stress exponent m decreases from $m = 40$ to $m = 1$. Reprinted by permission from Springer Nature: (Chang, Bulatov, and Yip 1999).

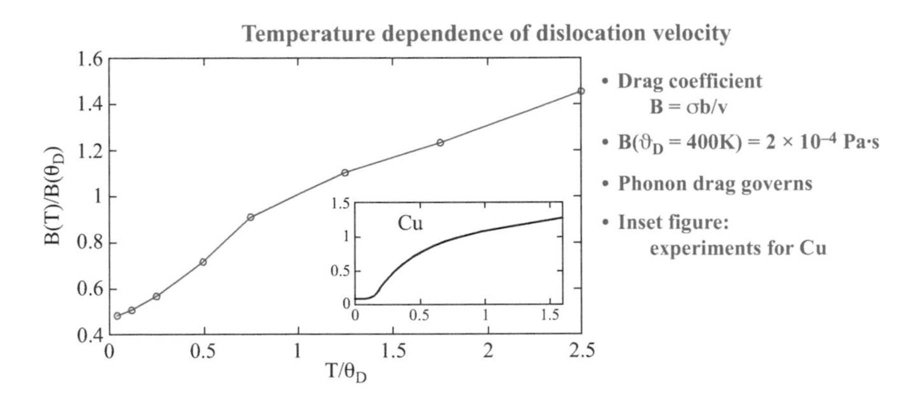

Figure 6.13
Temperature dependence of drag coefficient for edge dislocation glide in Mo. Simulation results are denoted by open circles. Inset shows experimental result of Cu (Jassby and Vreeland 1973). Reprinted by permission from Springer Nature: (Chang, Bulatov, and Yip 1999).

high-stress region, while experiments are generally confined to the low-velocity, low-stress region. In order to reach stresses comparable to the experimental range, one needs to resort to acceleration techniques that allow simulations over a considerably longer time period. This issue will be discussed in more depth in the essays in parts III and IV.

The dislocation drag coefficient B, $B = b\sigma/v$, is shown in figure 6.13, in normalized form, where θ_D is the Debye temperature. We see that the drag coefficient increases with increasing temperature, which means the velocity decreases with increasing temperature. This is fundamentally different from the behavior in the thermal activation region. The implication is that the underlying mechanism for the damping of edge dislocations is phonon drag. Although there are no experimental data on Mo to validate our interpretation, experiments on other materials do show a positive temperature dependence as illustrated in the inset figure. Theoretical calculations yield similar results (Alshits and Indenbom 1986). Another feature of the temperature dependence results is a plateau in the limit of low temperature, a behavior that is also suggested by the experiments.

Kink Mobility in Silicon

Because Si is a semiconductor material of great industrial interest, we describe a kinetic MC method of modeling the glide mobility of a single dislocation based on the sequence of elementary processes of double-kink nucleation, kink migration, and kink annihilation. The method is applied to study the coupling effects of the dissociated partial dislocations in Si, a system well known to have a high secondary Peierls barrier (Cai et al. 2000). As illustrated in figure 6.14, the PEL has undulations along two directions, the primary and secondary barrier heights are u_1 and u_2 respectively. The kink mechanism for a dislocation line, lying initially along one of the potential valleys, to move to the adjacent valley is to first nucleate a double kink (blue line) over the primary barrier, and then let this double kink expand across the secondary barrier so the dislocation line effectively slides over the primary barrier (in two steps). The entire process therefore involves two activation events, nucleation of a double kink of certain width and propagation of the left and right kinks (see figure 6.14), each with a characteristic energy that will have to be specified.

The kinetic MC method is a way to track the cumulative effects of nucleation anywhere along a dislocation line, and once a double kink is nucleated it can grow by kink migration, as illustrated schematically in figure 6.15. In this case, we have a dissociated dislocation described by a leading and a trailing partial dislocation bounding an area that is a stacking fault. The direction of motion is upward in the figure.

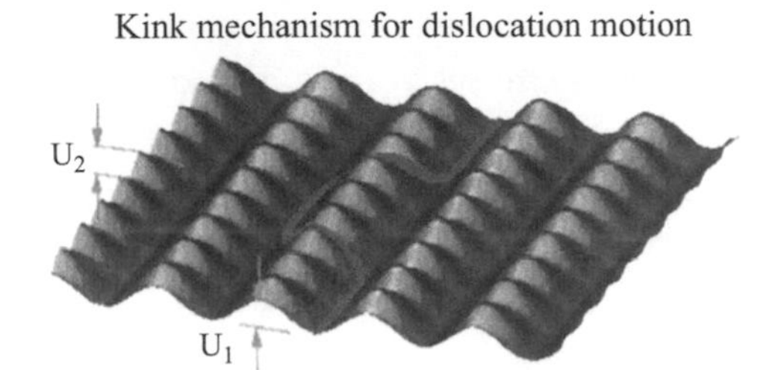

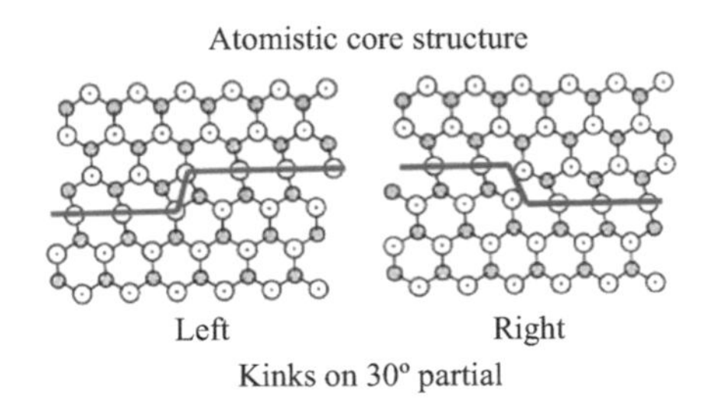

Figure 6.14
Schematic of Peierls barriers for double kink mechanism of dislocation motion (upper) and atomic configurations of left and right kinks on the 30° partial dislocation (lower). Figure from (Cai 2001).

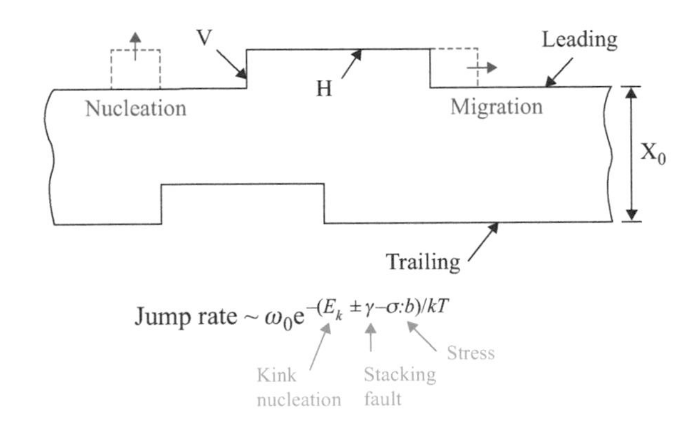

Figure 6.15
Schematic of dissociated dislocation (leading and trailing partials bounding a region of stacking fault. Along the leading partial nucleation and migration events are sampled according to a distribution which considers in addition to nucleation (or migration) activation the effects of the stacking fault energy and Peach–Koehler stresses. Figure from (Cai 2001).

The simulation is designed to produce the overall dislocation movement averaged over a large number of individual events. Required for input are the kink formation and migration energies, which in principle can be computed by electronic structure or atomistic calculations. The model we consider deals with a screw dislocation dissociated into two 30° partials in the (111) plane. Each partial is represented by a piecewise straight line composed of alternating horizontal (*H*) and vertical (*V*) segments. The length of *H* segments can be any multiple of the Burgers vector *b*, while the *V* segments all have the same length, the kink height *h*. The stacking fault bounded by the two parallel partials has a width measured in multiples of *h*. The simulation cell is oriented with the partials running horizontally so that a PBC can be applied in this direction.

In each simulation step, a stochastic sampling is performed to determine which event will take place next. The rates of these elementary events are calculated based on the energetics of the corresponding kink mechanisms. For example, for a kink pair nucleation three energy terms are considered, a formation energy, an energy bias favoring the reduction of the stacking-fault area, and the elastic interaction between a given segment and all the other segments and applied stress (so-called Peach–Koehler interaction). The estimate of the nucleation rate is what is shown in figure 6.15. A similar expression also exists for kink migration. The formation and migration energies used in the simulation are given in the table in figure 6.16, which shows the dislocation velocities obtained from the kMC simulations using two sets of atomistic input. While the results are seen to differ by some four orders of magnitude, they do bracket the experiments.

One could be cautiously optimistic about this comparison, considering the roughness of the calculation (kink multiplicity not considered and using rather approximate values for the frequency factor and entropy). Of the materials parameters that entered into the model, the kink energies remain uncertain despite efforts to obtain more accurate results using tight-binding and density functional theory methods. When more reliable atomistic results become available, we believe this approach can lead to significant understanding of the effects of applied stress on dislocation mobility in a system like Si.

Outlook for Multiscale Materials

It is safe to forecast that the computational materials community is currently facing science and technology challenges in innovation and design to meet the critical needs of our society, for example, materials in extreme environments, climate change, and

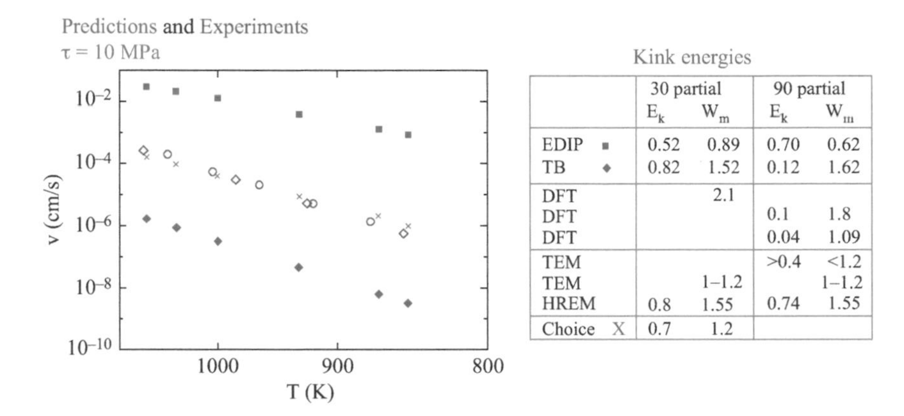

		30 partial		90 partial	
		E_k	W_m	E_k	W_m
EDIP	■	0.52	0.89	0.70	0.62
TB	◆	0.82	1.52	0.12	1.62
DFT			2.1		
DFT				0.1	1.8
DFT				0.04	1.09
TEM				>0.4	<1.2
TEM			1–1.2		1–1.2
HREM		0.8	1.55	0.74	1.55
Choice	X	0.7	1.2		

Figure 6.16
Temperature variation and magnitude of velocity of screw dislocation in Si, kMC results using two sets of kink nucleation and migration energies are denoted by square (EDIP potential) and closed diamond (tight binding), while experimental data are shown as red open diamond. Figure from (Cai 2001).

infrastructure sustainability. Besides understanding the kind of unit "process" problems that we have discussed, there will be needs to focus on larger-scale ("overarching") problems that can drive industrial-scale teams of investigators with complementary capabilities and shared interests. Virtual, microscale models will be needed to probe specific functional behavior of real materials on the mesoscale. In essence this is but another way of posing the challenge of linking nanoscale mechanics and physics with mesoscale and macroscale functionality and performance. Problems of this type come from application areas that can be very different from each other. Examples could include molecular-level understanding of the setting kinetics of cement (Lootens et al. 2004; Pellenq and Van Damme 2004), analyzing the fatigue resistance of bearing steel (Forst et al. 2006) using atomistic methods, and probing the oxidation resistance of an ultrahigh temperature ceramic by combining molecular simulation with experiments (Bongiorno et al. 2006). Further considerations of the outlook for computational materials will be discussed in the final section of the epilogue.

References

Alshits, V. I., and V. L. Indenbom. 1986. "Mechanisms of dislocation drag." *Dislocation in Solids* 7: 43.

Barrons, T. H. K., and M. L. Klein. 1965. "Second-order elastic constants of a solid under stress." *Proceedings of the Physics Society* 85: 523.

Basinski, Z. S., M. S. Duesbery, A. P. Pogany, R. Taylor, and Y. P. Varshni. 1970. "An effective ion–ion potential for sodium." *Canadian Journal of Physics* 48: 1480.

Bongiorno, A., C. J. Forst, R. K. Kalia, J. Li, J. Marschall, A. Nakano, M. M. Opeka, I. G. Talmy, P. Vashishsta, and S. Yip. 2006. "A perspective on modeling materials in extreme environments: oxidation of ultrahigh-temperature ceramics." *MRS Bulletin* 31: 410.

Born, M. 1940. "On the stability of crystal lattices. I." *Mathematical Proceedings of the Cambridge Philosophical Society* 36: 160.

Born, M., and K. Huang. 1956. *Dynamical Theory of Crystal Lattices*. Oxford: Clarendon.

Cai, W. 2001. "Atomistic and Mesoscale Modeling of Dislocation Mobility." PhD thesis, Massachusetts Institute of Technology, Department of Nuclear Engineering.

Cai, W., V. V. Bulatov, J. F. Justo, A. S. Argon, and S. Yip. 2000. "Intrinsic mobility of a dissociated dislocation in silicon." *Physical Review Letters* 15: 3346.

Chang, J. 2003. "Atomistics of Defect Nucleation and Mobility: Dislocations and Twinning." PhD thesis, Massachusetts Institute of Technology, Department of Nuclear Engineering.

Chang, J., V. V. Bulatov, and S. Yip, 1999. "Molecular dynamics study of edge dislocation motion in a bcc metal." *Journal of Computer-Aided Materials Design* 6: 165.

Chang, J., T. Zhu, J. Li, X. Lin, X. Qian, and S. Yip. 2004. "Multiscale Modeling of Defect Nucleation and Reaction: Bulk to Nanostructures." In *Mesoscopic Dynamics of Fracture Process and Materials Strength*, edited by H. Kitagawa and Y. Shibutani, 223. Dordrecht: Kluwer Academic.

Cleri, F., J. Wang, S. Yip. 1995. "Lattice instability analysis of a prototype intermetallic system under stress." *Journal of Applied Physics* 77: 1449.

Forst, C. J., J. Slycke, K. J. Van Vliet, and S. Yip. 2006. "Multiscale modeling of defect nucleation and reaction: bulk to nanostructures." *Physical Review Letters* 96: 175501.

Gouldstone, A., H.-J. Koh, K. Y. Zeng, A. E. Giannakopoulos, and S. Suresh. 2000. "Discrete and continuous deformation during nanoindentation of thin films." *Acta Materialia* 48: 2277.

Hill, R. 1962. "Acceleration waves in solids." *Journal of the Mechanics and Physics of Solids* 10: 1.

Hill, R. 1975. "On the elasticity and stability of perfect crystals at finite strain." *Mathematical Proceedings of the Cambridge Philosophical Society* 77: 225.

Hill, R., and F. Milstein. 1977. "Principles of stability analysis of ideal crystals." *Physical Review B* 15: 3087.

Hirth, J. P., and J. Lothe. 1982. *Theory of Dislocations*. New York: Wiley.

Hoover, W. G., A. C. Holt, and D. R. Squire. 1969. "Adiabatic elastic constants for argon. theory and Monte Carlo calculations." *Physica* 44: 437.

Jassby, K. M., and T. Vreeland. 1973. "Dislocation mobility in pure copper at 4.2 K." *Physical Review B* 8: 3537.

Jonsson, H., G. Mills, and K. W. Jacobsen. 1998. "Nudged elastic band method for finding minimum energy paths of transition." In *Classical and Quantum Dynamics in Condensed Phase Simulations*, edited by B. J. Berne, G. Ciccotti, and D. F. Coker, 385. New York: Plenum Press.

Kelly, A., and N. H. Macmillan. 1986. *Strong Solids*, 3rd ed. Oxford: Clarendon.

Klahn, D., A. K. Mukherjee, and J. E. Dorn. 1970. "Strain-rate effects." *Strength of Metals and Alloys, Proceedings of the Second International Conference*, p. 951. Cleveland, OH: American Society of Metals.

Langely, W. R. and R. G. Van Name, eds. 1928. *The Collected Works of J. Willard Gibbs*. New York: Longmans, Green and Co.

Li, J. 2000. "Modeling Microstructural Effects of Deformation Resistance and Thermal Conductivity." PhD thesis, Massachusetts Institute of Technology, Dept. of Nuclear Engineering.

Li, J., K. J. Van Vliet, T. Zhu, S. Yip, and S. Suresh. 2002. "Atomistic mechanisms governing elastic limit and incipient plasticity in crystals." *Nature* 418: 307.

Li, J., and S. Yip. 2002. "Atomistic measures of materials strength." *Computer Modelling in Engineering and Sciences* 3: 219.

Li, J., W. Cai, J. Chang, and S. Yip. 2003. "Computational materials science." *NATO Science Series Sub Series III Computer And Systems Sciences* 187: 359–387.

Li, J., T. Zhu, S. Yip, K. Van Vliet, and S. Suresh. 2004. "Elastic criterion for dislocation nucleation." *Materials Science and Engineering. A* 365: 25.

Lootens, D., P. Hebraud, E. Lecolier, and H. van Damme. 2004. "Gelation, shear-thinning and shear-thickening in cement slurries." *Oil & Gas Science Technology—Review IFP* 59: 31.

Mizushima, K., S. Yip, and E. Kaxiras. 1994. "Ideal crystal stability and pressure-induced phase transition in silicon." *Physical Review B* 50: 14952.

Morris, J. W., and C. R. Krenn. 2000. "The internal stability of an elastic solid." *Philosophical Magazine A* 80: 2827.

Morris, J. W., C. R. Krenn, D. Roundy, and M. L. Cohen. 2000. "Elastic Stability and the Limit of Strength." In *Phase Transformations and Evolution in Materials*, edited by P. E. Turchi and A. Gonis, 187. Warrendale, PA: TMS.

Nadgornyi, E. 1988. "Dislocation dynamics and mechanical properties of crystals." In *Progress in Materials Science*, Vol. 31.

Neuhauser, H. 1979. "Dynamical effects in deformation." In *Strength of Metals and Alloys, Proceedings of the Fifth International Conference*, edited by P. Haasen, V. Gerold, and G. Kostorz, 1531. Oxford: Pergamon Press.

Parrinello, M., and A. Rahman. 1981. "Polymorphic transitions in single crystals: a new molecular dynamics method." *Journal of Applied Physics* 52: 7182.

Pellenq, R. J.-M., and H. van Damme. 2004. "Why does concrete set?: The nature of cohesion forces in hardened cement-based materials." *MRS Bulletin* 29: 319.

Rice, J. R. 1976. "The localization of plastic deformation." In *Theoretical and Applied Mechanics*, edited by W. T. Koiter, Vol. 1, 207. Amsterdam: North-Holland.

Tang, M., and S. Yip. 1994. "Lattice instability in β-SiC and simulation of brittle fracture." *Journal of Applied Physics* 76: 2719.

Tang, M., and S. Yip. 1995. "Lattice instability in β-SiC and simulation of brittle fracture." *Physical Review Letters* 75: 2738.

Tersoff, J. 1989. "Modeling solid-state chemistry: interatomic potentials for multicomponent systems." *Physical Review B* 39: 5566.

Van Vliet, K. J., J. Li, T. Zhu, S. Yip, and S. Suresh. 2003. "Quantifying the early stages of plasticity through nanoscale experiments and simulations." *Physical Review B* 67: 104105.

Wallace, D. C. 1972. *Thermodynamics of Crystals*. New York: Wiley.

Wang, J., S. Yip, S. Phillpot, and D. Wolf. 1993. "Crystal instabilities at finite strain." *Physical Review Letters* 71: 4182.

Wang, J., J., Li, S. Yip, S. Phillpot, and D. Wolf. 1995. "Mechanical instabilities of homogeneous crystals." *Physical Review B* 52: 12627.

Yip, S. 2003. "Synergistic science." *Nature Materials* 2: 3.

Yip, S. 2006. "Soft-mode scenarios of shear localization: atomic-level landscapes." *Journal of Statistics and Physics* 125: 1113.

Zhou, Z., and B. Joos. 1996. "Stability criteria for homogeneously stressed materials and the calculation of elastic constants." *Physical Review B* 54: 3841.

Zhu, T., J. Li, and S. Yip. 2004a. "Atomistic study of dislocation loop emission from a crack tip." *Physical Review Letters* 93: 025503.

Zhu, T., J. Li, and S. Yip. 2004b. "Atomistic configurations and energetics of crack extension in silicon." *Physical Review Letters* 93: 205504.

Zhu, T., J. Li, K. J. Van Vliet, S. Suresh, S. Ogata, and S. Yip. 2004. "Predictive modeling of nanoindentation-induced homogeneous dislocation nucleation in copper." *Journal of the Mechanics and Physics of Solids* 52: 691.

Further Reading

Bulatov, V. V., S. Yip, and A. S. Argon. 1995. "Atomic modes of dislocation mobility in Silicon." *Philosophical Magazine A* 72: 453.

Bulatov, V., F. F. Abraham, L. Kubin, B. Devincre, and S. Yip. 1998. "Connecting atomistic and mesoscale simulations of crystal plasticity." *Nature* 391: 669.

Justo, J. F., M. Z. Bazant, E. Kaxiras, V. V. Bulatov, S. Yip. 1998. "Interatomic potential for silicon defects and disordered phases." *Physical Review B* 58: 2539.

Ngadi, A., and J. Rajehenbach. 1998. "Intermittencies in the compression of a model granular medium." *Physical Review Letters* 80: 273. Observations of large sudden fluctuations in the force chains acting in confined granular systems.

Cai, W., V. V. Bulatov, J. Chang, J. Li, and S. Yip. 2001. "Anisotropic elastic interactions of a periodic disocation array." *Physical Review Letters* 86: 5727.

Chang, J., W. Cai, V. V. Bulatov, S. Yip. 2001. "Dislocation motions in bcc metals by molecular dynamics." *Materials Science and Engineering. A* 309–310: 160.

Cai, W., V. V. Bulatov, S. Yip, and A. S. Argon. 2001. "Kinetic Monte Carlo modeling of dislocation motion in bcc metals." *Materials Science and Engineering. A* 309–310: 270.

George, A., and S. Yip. 2001. "Preface to the Viewpoint Set on: dislocation mobility in silicon." *Scripta Materialia* 45: 1233.

Yip, S., J. Li, M. Tang, and J. Wang. 2001. "Mechanistic aspects and atomic-level consequences of elastic instabilities in homogeneous crystals." *Materials Science and Engineering. A* 317: 236. Elastic stability criteria for a homogeneous lattice under arbitrary external loading combined with MD simulation allow a systematic analysis of competing structural transitions as in pressure-induced polymorphic and crystal-to-amorphous transitions.

Becker, R. 2002. "Developments and trends in continuum plasticity." *Journal of Computer-Aided Materials Design* 9: 165–172. Focused on multiscale modeling, anisotropy effects of grain deformation, dynamic response of shear localization, twinning and phase transformation models.

Moriarty, J. A., Vitek, V. V. Bulatov, and S. Yip. 2002. "Atomistic simulations of dislocations and defects." *Journal of Computer-Aided Materials Design* 9: 99–132. Multiscale modeling of plasticity and strength with emphasis on bcc metals and deformation at extreme conditions.

Cai, W., V. V. Bulatov, J. Chang, J. Li, and S. Yip. 2003. "Dislocation Core Effects on Mobility." In *Dislocations in Solids*, edited by F. N. R. Nabarro and J. P. Hirth, Vol. 13. Amsterdam: Elsevier.

Li, J., W. Cai, J.-P. Chang, and S. Yip. 2003. "Atomistic Measures of Materials Strength and Deformation." In *Computational Materials Science*, edited by R. Catlow and E. Kotomin, 359. Amsterdam: IOS Press.

Lu, J., G. Ravichandran, and W. L. Johnson. 2003. "Deformation behavior of the $Zr_{41.2}Ti_{13.8}Cu_{12.5}Ni_{10}Be_{22.5}$ bulk metallic glass over a wide range of strain-rates and temperatures."

Acta Materialia 51: 3429. Stress and viscosity data allow a mechanism map in strain rate and temperature showing regimes of Newtonian, non-Newtonian flows, and shear localization.

Li, J., T. Zhu, S. Yip, K. Van Vliet, and S. Suresh. "Elastic criterion for dislocation nucleation." *Materials Science and Engineering. A* 365: 25. Notion of theoretical strength is extended to nonuniform loading and coupled to MD and finite-element simulation to derive a local stiffness measure for modeling incipient plasticity in a thin film material.

Shi, Y., and M. L. Falk. 2005. "Strain localization and percolation of stable structure in amorphous solids." *Physical Review Letters* 95: 095502. MD simulation study of the effect of structural relaxation prior to mechanical testing, with more rapidly quenched initial structures undergoing more localization.

Ogata, S., J. Li, and S. Yip. 2005. "Energy landscape of deformation twinning in bcc and fcc metals." *Physical Review B* 71: 224102. Density functional theory calculations to explore the condition for runaway defect growth in the case of thickness of the twin regions.

Silva, E. C. C. M., L. Tong, S. Yip, and K. J. Van Vliet. 2006. "Size effects on the stiffness of silica nanowires." *Small* 2: 395.

Foerst, C. J., J. Slcke, K. J. Van Vliet, and S. Yip. 2006. "Point defect concentrations in metastable Fe-C alloys." *Physical Review Letters* 96: 175501.

Bongiorno, A., C. J. Forst, R. K. Kalia, J. Li, J. Marschall, A. Nakano, M. M. Opeka, I. G. Talmy, P. Vashishsta, and S. Yip. 2006. "A perspective on modeling materials in extreme environments: oxidation of ultra-high temperature ceramics." *Materials Research Society Bulletin* 31: 410.

Wang, C.-Z., J. Li, K.-M. Ho, and S. Yip. 2006. "Undissociated screw dislocation in Si: glide or shuffle set?" *Applied Physics Letters* 89: 051910.

Suresh, S. 2006. "Colloid model for atoms." *Nature Materials* 5: 253.

Schuh, C. A. 2006. "Nanoindentation studies of materials." *Materials Today* 9 (5): 32. Discrete events including dislocation source activation, shear instability initiation, and phase transformations can be detected with high resolution and in a dynamical rate-sensitive mode.

Izumi, S., and S. Yip. 2008. "Dislocation nucleation from a sharp corner in silicon." *Journal of Applied Physics* 104: 033513.

Shan, Z. W., J. Li, Y. Q. Cheng, A. M. Minor, S. A. Syed Asif, O. L. Warren, and E. Ma. 2008. "Plastic flow and failure resistance of metallic glass: insight from *in situ* compression." *Physical Review B* 77: 155419. Suppression of shear localization by confinement and sample-size effects explored experimentally and by scaling analysis.

Yip, S. 2009. "Multiscale Materials." In *Multiscale Methods*, edited by J. Fish, 481. New York: Oxford Univ. Press.

Zhu, T., J. Li, S. Ogata, and S. Yip. 2009. "Mechanics of ultra-strength materials." *MRS Bulletin* 34: 167.

Kabir, M., T. T. Lau, D. Rodney, S. Yip, and K. J. Van Vliet. 2010. "Predicting dislocation climb and creep from explicit atomistic details." *Physical Review Letters* 105: 095501.

Kabir, M., T. T. Lau, X. Lin, S. Yip, and K. J. Van Vliet. 2010. "Effects of vacancy-solute point defect clusters on diffusivity in metastable Fe-C alloys." *Physical Review B* 82: 134112.

Langer, J. S., E. Bouchbinder, and T. Lookman. 2010. "Thermodynamic theory of dislocation-mediated plasticity." *Acta Materialia* 58: 3718. Early reformulation of theory of polycrystalline plasticity under nonequilibrium, externally driven conditions treating flow of energy and entropy associated with dislocations.

Zhu, T., J. Li, and S. Yip. 2013. "Atomistic reaction pathway sampling: The nudged elastic band method and nanomechanics applications." In *Nano and Cell Mechanics: Fundamentals and Frontiers*, edited by H. Espinosa and G. Bao, chap. 12, 313. Chichester, UK: Wiley.

Guo, Y. F., S. Xu, X.-Z. Tang, Y.-S. Wang, S. Yip. 2014. "Twinnability of hcp metals at the nanoscale." *Journal of Applied Physics* 115: 224902.

Ding, J., S. Patinet, M. L. Falk, Y. Cheng, and E. Ma. 2014. "Soft spots and their structural signature in a metallic gas." *Proceedings of the National Academy of Science* 111: 14052. Demonstrates correlation between soft vibrational modes associated with the most disordered local polyhedral packing environments and shear transformation zones composed of atoms with large nonaffine displacements, effectively a statistical coupling of topological heterogeneity with vibrational-relaxational heterogeneity.

Wang, J., Z. Zeng, C. R. Weinberger, Z. Zhang, T. Zhu, and S. X. Mao. 2015. "*In situ* atomic-scale observation of twinning-dominated deformation in nanoscale body-centered cubic tungsten." *Nature Materials* 14: 594. Demonstration and rationalization of deformation twinning as a mechanism in bcc nanostructures.

Cui, Y., G. Po, and N. Ghoniem. 2016. "Controlling strain bursts and avalanches at the nano- to micrometer scale." *Physical Review Letters* 117: 155502. Dislocation dynamics simulations show that strain bursts have scale-free avalanche statistics similar to critical phenomena in general.

7 Ideal Shear Strength

Although aluminum (Al) has a smaller shear modulus than copper, one finds by electronic-structure calculations that its ideal shear strength is actually larger by virtue of a more extended deformation range. This fundamental difference can be explained by examining the valence charge redistributions as each metal is subjected to shear. The charge density in Al is strongly directionally dependent relative to copper, which in turn accounts for its abnormally high intrinsic stacking fault energy and significant orientation dependence in pressure hardening.

Incorporating such notions in the concept of *shearability* of matter, we find a gap between two classes of materials, metals and ceramics, as well as further insights into the effects of valence charge distributions on the behavior of solids in shear. On the one hand, shearability of metals correlates with the degree of charge localization and directional bonding. On the other hand, depending on deformation constraints, ionic solids may possess even greater shearability than covalent solids. Overall, the Frenkel model of ideal shear strength is applicable to both metals and ceramics when shearability is used as a scaling parameter.

Introduction

The minimum shear stress necessary to cause a permanent deformation in a pristine crystal (without imperfections) is fundamental to understanding materials strength and its theoretical limits at large strain (Wang et al. 1995; Morris and Krenn 2000). With the possible exceptions of nanoindention measurements (Gouldstone et al. 2000), it has not been feasible to directly measure the ideal shear strength of crystals. The demonstration that this property can be reliably determined by first-principles calculations therefore would have implications for controlling the behavior of solids at the limit of structural stability. Results on stress-strain behavior of Al and Cu in $(111) < 11\bar{2} >$ shear, calculated with density functional theory (DFT) and accounting for full atomic relaxation, initially showed Cu having a higher ideal shear strength than Al (Roundy et al. 1999). However, a later study, also using DFT methods, instead found Al has the higher strength (Ogata, Li, and Yip 2002). The latter is based on a more extensive analysis of the energetics of shear deformation, the pressure hardening behavior, and valence charge distribution during deformation. These results show that the ideal shear strength and related properties, such as stacking-fault energies, are sensitive to the underlying electronic structure. Specifically, bonding in Al is much more like a "hinged rod" during the breaking and reformation of directional bonds, whereas the behavior in Cu is more isotropic, like "sphere-in-glue."

Multiplane Analysis

The intrinsic stacking-fault energy is a measure of the energy penalty when two adjacent planes in a crystalline lattice are sheared relative to each other. It is known to play an important role in the structure and energetics of dislocations formed by slip processes. While experiments already can tell us the intrinsic stacking-fault energy is much larger in Al than in Cu, this knowledge by itself is not sufficient to fully appreciate the ideal shear strength of crystals. One needs a broader theoretical framework in which the stacking fault energy appears as a check point.

Consider the energy function

$$\gamma_n(x) \equiv \frac{E_n(x)}{nS_0},\ n = 1, 2, \ldots \tag{7.1}$$

where x is the relative displacement in the slip direction between two adjacent atomic planes (we focus on $\{111\}\langle 11\bar{2}\rangle$ slip here), $E_n(x)$ is the increase in total energy relative to its value at $x = 0$, with $n+1$ being the number of planes involved in the shearing and S_0 being the cross-sectional area at $x = 0$. The series of functions $\gamma_1(x)$, $\gamma_2(x)$, . . . , $\gamma_\infty(x)$ may be called the multiplane generalized stacking fault energies, with $\gamma_1(x)$ being the conventional generalized stacking fault energy (GSF) (Zimmerman, Gao, and Abraham 2000) and $\gamma_\infty(x)$ the affine strain energy (see figure 7.1).

The intrinsic stacking fault energy γ_{sf} is $\gamma_1(b_p)$, where $\vec{b}_P = [11\bar{2}]a_0/6$ is the partial Burgers vector and a_0 the equilibrium lattice constant. The unstable stacking energy γ_{us}, an important parameter in determining the ductility of the material (Rice and Beltz 1994), is $\gamma_1(x_0)$, where $d\gamma_1/dx(x_0 < b_P) = 0$. The asymptotic behavior at large n can be expressed as

$$\gamma_n(x) = \gamma_\infty(x) + \frac{2\gamma_{twin}(x)}{n} + O(n^{-2}) \tag{7.2}$$

where $\gamma_{twin}(b_p)$ is the unrelaxed twin boundary energy. The rate of convergence to equation (7.2) reflects the localization range of metallic bonding in a highly deformed bulk environment.

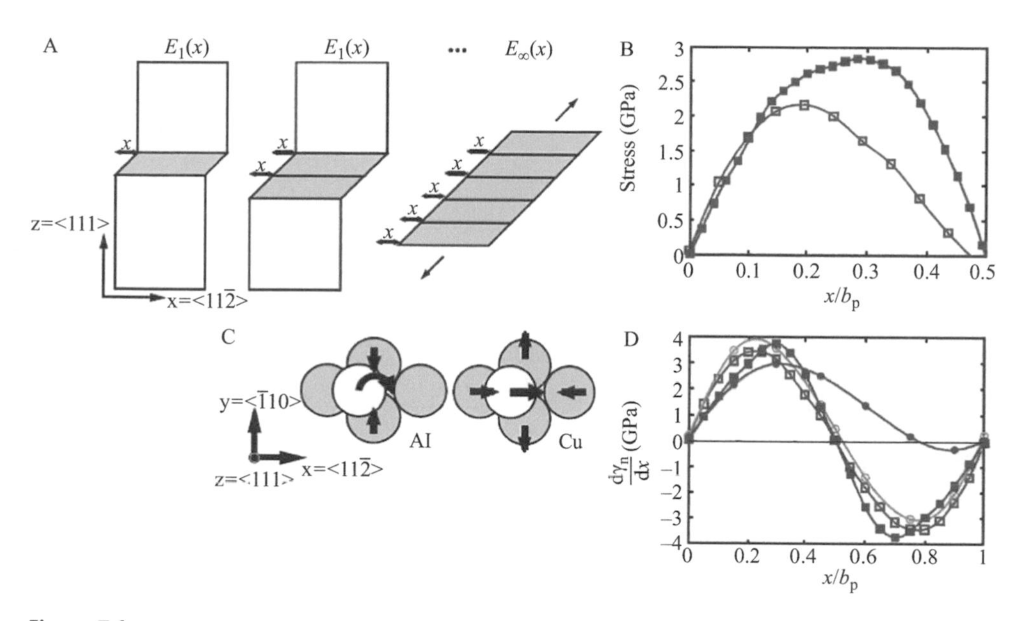

Figure 7.1
(a) Crystal configuration for shearing of one, two, . . . ∞ planes by a displacement x in the orientation indicated with corresponding multiplane generalized stacking fault energies. (b) Stress-displacement responses in pure shear of Al (solid squares) and Cu (open squares). (c) Ion relaxation patterns in Al and Cu. (d) Comparison of stress-displacement responses $d\gamma_\infty(x)/dx$ (squares) and $d\gamma_1(x)/dx$ (circles) in Al (solid symbols) and Cu (open symbols). Figure used with permission of American Association for the Advancement of Science, from (Ogata, Li, and Yip 2002); permission conveyed through Copyright Clearance Center, Inc.

The DFT calculations, following the same procedure previously described (Roundy et al. 1999), are shown in table 7.1. Further details are given in the original report (Ogata, Li, and Yip 2002). Notice the results show Al has a much larger γ_{sf} value than Cu, yet their γ_{us} values are quite close.

At equilibrium Cu is considerably stiffer than Al; its bulk, simple, and pure shear—along (111)(11$\bar{2}$)—moduli are greater than those of Al by 80, 65, and 25%, respectively. However, Al has an ideal pure shear strength that is 32% larger than that of Cu. The explanation, provided by the present results, points to Al having a longer range of elastic strain before softening. Notice in figure 7.1(b) the displacement and the engineering shear strain at the maximum shear stress in Al is $x_{max}/b_p = 0.28$ and $\gamma_{max} = 0.20$, compared with $x_{max}/b_p = 0.19$ and $\gamma_{max} = 0.13$ in Cu. Moreover, the ion relaxations in these two metals are different as shown in figure 7.1(c). In Al, when the top atom slides over the bottom atoms, it also hops in the z direction while the bottom atoms contract in the y direction (and essentially no relaxation in x). In contrast, in Cu there is almost

Table 7.1
Benchmark results, comparison of present calculations (Calc), experiments (Expt), and previous calculations (Oth calc). Dashes indicate results are not available (Ogata, Li, and Yip 2002).

	Al			Cu		
Variable	Calc[a]	Expt	Oth calc	Calc[a]	Expt	Oth calc
a_0(Å)	4.04	4.03[b]	4.04[c]	3.64	3.62[b]	3.64[d]
G'_r (GPa)	25.4	27.4[e]	19–25[f]	31.0	33.3[e]	26–34[f]
G'_u (GPa)	25.4	27.6[e]	24–30[f]	40.9	44.4[e]	36–44[f]
γ_{sf} (mJ/m^2)	158	166[g]	143[h], 164[i]	39	45[g]	(49)[j]
γ_{us} (mJ/m^2)	175	—	183[h], 224[i]	158	—	(210)[j]

[a] VASP, US-GGA, $18\times25\times11$ Monkhorst-Pack $\vec{k}$ points.
[b] Al at temperature $T=0$ K, Cu at $T=298$ K (Hellwege and Hellwege 1988).
[c] GGA (Stampfl and Van de Walle 1999).
[d] Full-potential linearized augmented plane wave method (WIEN97 program), GGA (Jona and Marcus 2001).
[e] calculated from elastic constants at $T=0$ K (Nelson 1992).
[f] LDA (Roundy et al. 1999).
[g] (Hirth and Lothe 1982).
[h] LDA (Hartford et al. 1998).
[i] LDA (Lu et al. 2000).
[j] LDA, unrelaxed (Zimmerman, Gao, and Abraham 2000).

no relaxation in the z direction as the top atom translates horizontally, and the bottom atoms expand and contract in the y and x directions, respectively.

The difference in relaxation patterns has important implications for the shear strength–hardening behavior which has been noted previously and discussed in terms of third-order elastic constant (Krenn et al. 2001). When pressurized in the (110) direction, Cu hardens while Al softens substantially. However, if pressurized in the (111) direction, Al hardens substantially, while Cu softens slightly. These results show the pressure-hardening is a highly orientation- dependent effect.

The charge-density behavior just discussed, along with the relaxation patterns seen in figure 7.1(c), suggest a hinged-rod model to describe the shear strength for Al, in contrast to the conventional "muffin-tin" or sphere-in-glue model for Cu. It is reasonable to think that when the bonding is directionally dependent (rod-like), a longer range of deformation can be sustained before breaking than when the bonding is spherically symmetric. In covalent systems like Si (Umeno and Kitamura 2002) and SiC, it has been verified that during shear, the bonds generally do not break until the engineering shear strain reaches 25–35%, substantially larger than those of metallic systems. Conversely, when the bonds do break, a directionally bonded system can be expected to be more frustrated and less accommodating, giving rise to a larger intrinsic stacking fault energy.

To quantify the foregoing intuitive reasoning, one can look to the behavior of the multiplane generalized stacking fault energies in the form of stress-displacement functions $d\gamma_1(x)/dx$ and $d\gamma_\infty(x)/dx$, figure 7.1(d). For Cu $d\gamma_1(x)/dx$ and $d\gamma_\infty(x)/dx$ are not very different across the entire range of shear, so the sphere-in-glue picture is appropriate. The fact that the sliding of a layer is effectively decoupled from the adjacent layers is consistent with the bonding in Cu, which is essentially bond-angle independent. The situation is quite different in Al, especially in the strain range $x > x_{max}$, where the gradient reaches a maximum. Even in the range of $x < x_{max}$, the relative magnitudes of $d\gamma_1(x)/dx$ and $d\gamma_\infty(x)/dx$ are opposite in order in Al as compared to those in Cu, thus indicating a decidedly different type of orientational bonding. Secondly, the value of x_{max} is almost identical between $d\gamma_1(x)/dx$ and $d\gamma_\infty(x)/dx$ in both Al and Cu, with Al having the larger x_{max} value, implying the longer range directional bonding in Al could be a more general feature than being specific only to the affine strain energy $\gamma_\infty(x)$. Lastly, when $x \gg x_{max}$ and the directional bonds in Al are broken, $d\gamma_1(x)/dx$ in Al stays positive for an extended range, whereas $d\gamma_1(x)/dx$ in Cu becomes negative quickly. Notice that Al and Cu have approximately the same unstable stacking energy (table 7.1). When the displacement x reaches b_P and the configuration becomes an intrinsic stacking fault, Cu has recovered most of its losses in the sense of a low value of γ_{sf}, while Al has recovered very little as its γ_{sf} value remains close to the γ_{us} value. Thus, when a directional bond is broken it is more difficult for the electrons to redistribute. For sphere-in-glue–type systems, even if the bond angles are wrong, so long as the volumes are not highly nonuniform (as in the intrinsic stacking fault), the electrons will readily redistribute themselves. Under such conditions, the system should not incur a large energy penalty.

In this essay, the connection between the generalized stacking fault energy and the stress-strain response is used to explain the abnormally high ideal shear strength and intrinsic stacking fault energy of Al stem from the same electronic-structure origin. On the one hand, directional bonds give rise to a relatively longer shear-deformation range, which accounts for the larger ideal shear strength of Al relative to Cu. On the other hand, once the existing bonds are broken and new bonds are formed with unfavorable bond angles, the electrons cannot readjust easily, resulting in an anomalous intrinsic stacking fault energy for Al. These findings are supported by the detailed behavior of the valence charge density obtained from first-principles calculations. Furthermore, they suggest that conventional crystal plasticity notions such as a scalar or pressure-independent yield criterion based on critical resolved shear stress (although successful for macroscopic face-centered cubic metals) should be viewed with caution when interpreting

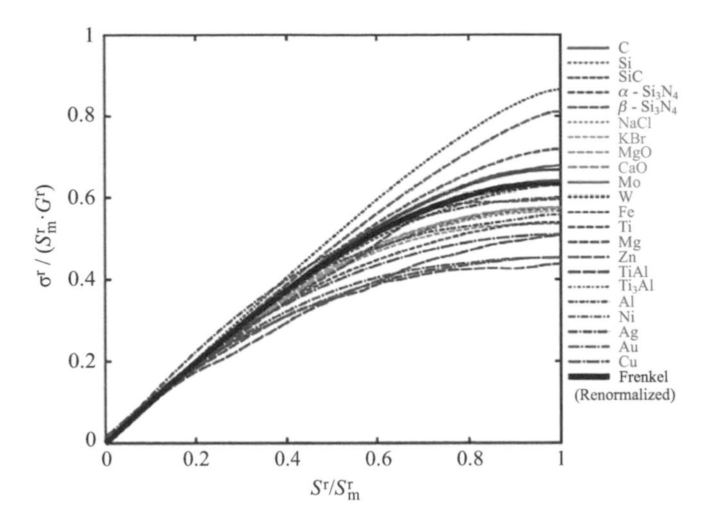

Figure 7.2
Normalized stress-strain behavior of metals and ceramics computed by density-functional theory methods showing shear softening after an initial elastic behavior. Also shown is the Frenkel model modified to have an initial slope (normalized) of $2/\pi$ instead of unity. Reprinted figure with permission from (Ogata et al. 2004). Copyright 2004 by American Physical Society.

nanoindentation experiments (Gouldstone et al. 2000). Contemporary empirical potentials (Mishin et al. 2001) may be useful for providing a qualitative description of the nonlinear, anisotropic stress distribution under the nanoindenter and for ascertaining the likely site and character of the instability; the quantitative implications of these results remain to be scrutinized by more accurate ab initio calculations.

Shearability of Metals and Ceramics

Fundamental understanding of *ductility* of solids begins with the notion of comparing the energies needed to break an interatomic bond by shear and by tension (Frenkel 1926; Peierls 1940; Nabarro 1947; Foreman, Jaswon, and Wood 1951; Rice 1992; Rose, Smith, and Ferrante 1983; Xu, Argon, and Ortiz 1995). Using density functional theory methods, the stress-strain response of metals and ceramics have been computed to determine the maximum shear stress a homogeneous crystal can withstand, a property that we will denote as *shearability*. Among the metals and covalent ceramics analyzed, and considering both relaxed and unrelaxed atomic configurations, a *ductility* gap is revealed, as shown in figure 7.2 (Ogata et al. 2004).

Figure 7.2 is a stress-strain plot for strains up to the point where the stress is a maximum, normalized such that all curves have initial unit slope and reach maximum stress when the reduced strain is at unity. This scaling allows the results for 22 metals and ceramics to collapse onto a master curve. Also indicated is the Frenkel model of a quarter sinusoid

$$\sigma = \frac{2Gs_m}{\pi}\sin\left(\frac{\pi s}{2s_m}\right),\quad 0 < s < s_m,\quad \sigma_m = \frac{2Gs_m}{\pi} \tag{7.3}$$

which gives an initial slope of $2/\pi$. This simple model serves as a rough overall description of the group of metals and ceramics explicitly considered. Notice equation (7.3) differs from the original Frenkel model that gives an initial slope of unity (Frenkel 1926). Put another way, equation (7.3) described a universal shear-softening response which is the counterpart to the Universal Binding Energy Relation for deformation in uniaxial tension (Rose 1983). This is also a two-parameter relation which has been checked against DFT calculations. Taken together they allow materials design and performance criteria to be formulated in which tensile and shear dissipation modes compete (Rice 1992; Xu, Argon, and Ortiz 1995; Kysar 2003; Vitos, Korzhavyi, and Johansson 2003). The existence of the gap can also be interpreted by a mechanism map of ideal shear and tensile strengths, shown in figure 7.3.

Implications

The ratio of shear to bulk modulus, G/B, is accessible experimentally and has been used as a performance predictor in alloy design, for example, to predict the optimum compromise between ductility and corrosion resistance in stainless steels (Vitos, Korzhavyi, and Johansson 2003). On the other hand, maximum shear and tensile strains, while feasible to obtain by ab initio calculations, are unavailable experimentally, and therefore have not been used in a practical manner, despite their established conceptual importance (Foreman, Jaswon, and Wood 1951; Xu, Argon, and Ortiz 1995). The existence of the gap in the ratio of γ_{us}/γ_s signifies a dependence not only on the crystal structure, but also on the bonding and the loading conditions. Moreover, the maximum tensile strain shows less sensitivity than the maximum shear strain (Ogata et al. 2004). Figure 7.3 is a way to summarize the present findings. It also reminds us that reliable electronic calculations can play two distinct roles in probing molecular mechanisms—an *explanatory* role such as understanding why Al has a larger ideal shear strength than Cu, and a *discovery* or *predictive* role such as finding a shearability gap between metals and ceramics.

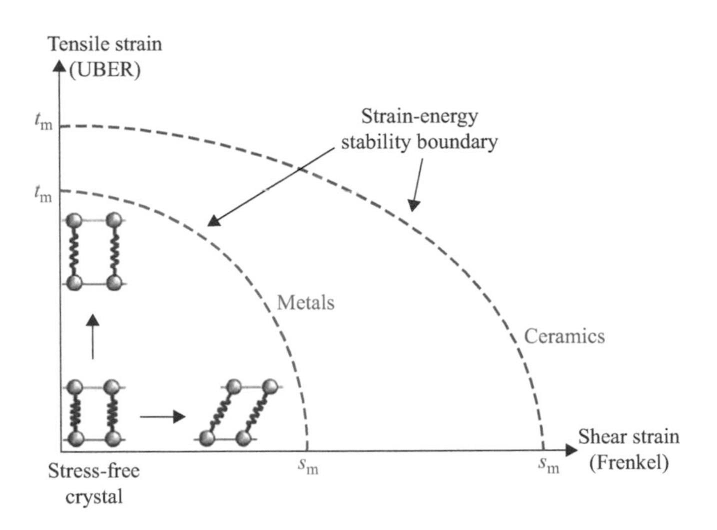

Figure 7.3
Correlation between tensile and shear strains predicted by density-functional theory calculations for metals and ceramics. Tensile strains are based on the model UBER (Rose, Smith, and Ferrante 1983), whereas shear strains are based on the Frenkel model, equation (7.3) (Frenkel 1926). A gap in shearability is clearly illustrated in this mechanism map. Reprinted figure with permission from (Ogata et al. 2004). Copyright 2004 by American Physical Society.

References

Foreman, A. J., M. A. Jaswon, and J. K. Wood. 1951. "Factors controlling dislocation widths." *Proceedings of the Physical Society of London, Section A* 64: 156.

Frenkel, J. 1926. "Zur Theorie der Elastizitätsgrenze und der Festigkeit kristallinischer Körper." *Zeitschrift für Physiks* 37: 572.

Gouldstone, A., H. J. Koh, K. Y. Zeng, A. E. Giannakopoulos, and S. Suresh. 2000. "Discrete and continuous deformation during nanoindentation of thin films." *Acta Materialia* 48: 2277.

Hartford, J., B. von Sydow, G. Wahnstrôm, and B. I. Lundqvist. 1998. "Peierls barriers and stresses for edge dislocations in Pd and Al calculated from first principles." *Physical Review* B 58: 2487.

Hellwege, K.-H., and A. M. Hellwege (eds.). 1988. *Structure Data of Elements and Intermetallic Phases. Elements, Borides,* Landolt-Bôrnstein III/14a. New York: Springer-Verlag.

Hirth, J. P., and J. Lothe. 1982. *Theory of Dislocations*, 2nd ed. New York: Wiley.

Jona, F., and P. M. Marcus. 2001. "Structural properties of copper." *Physical Review* 8 (63): 094113.

Krenn, C. R., D. Roundy, J. W. Morris Jr., and M. L. Cohen. 2001. "The non-linear elastic behavior and ideal shear strength of Al and Cu." *Materials Science and Engineering. A* 317: 44.

Kysar, J. W. 2003. "Energy dissipation mechanisms in ductile fracture." *Journal of the Mechanics and Physics of Solids* 51: 795.

Lu, G., N. Kioussis, V. V. Bulatov, and E. Kaxiras. 2000. "Generalized-stacking-fault energy surface and dislocation properties of aluminum." *Physical Review B* 62: 3099.

Mishin, Y., M. J. Mehl, D. A. Papaconstantopoulos, A. F. Voter, and J. D. Kress. 2001. "Structural stability and lattice defects in copper: ab initio, tight-binding, and embedded-atom calculations." *Physical Review B* 63: 224106.

Morris, J. W. Jr., and C. R. Krenn. 2000. "The internal stability of an elastic solid." *Philosophical Magazine A* 80: 2827.

Nabarro, F. R. N. 1947. "Dislocations in a simple cubic lattice." *Proceedings of the Physics Society of London* 59: 256.

Nelson, D. F. (ed.). 1992. *Low Frequency Properties of Dielectric Crystals Second and Higher Order Elastic Constants,* Landolt-Bôrnstein III/29a. New York: Springer-Verlag.

Ogata, S., J. Li, and S. Yip. 2002. "Ideal pure shear strength of aluminum and copper." *Science* 298: 807.

Ogata, S., J. Li, N. Hirosaki, Y. Shibutani, and S. Yip. 2004. "Ideal shear strain of metals and ceramics." *Physical Review B* 70: 104104.

Peierls, R. 1940. "The size of a dislocation." *Proceedings of the Physics Society of London* 52: 34.

Rice, J. R. 1992. "Dislocation nucleation from a crack tip: an analysis based on the Peierls concept." *Journal of the Mechanics and Physics of Solids* 40: 239.

Rice, J. R., and G. E. Beltz. 1994. "The activation energy for dislocation nucleation at a crack." *Journal of the Mechanics and Physics of Solids* 42: 333.

Rose, J. H., J. R. Smith, and J. Ferrante. 1983. "Universal features of bonding in metals." *Physical Review B* 28: 1835.

Roundy, D., C. R. Krenn, M. L. Cohen, and J. W. Morris. 1999. "Ideal shear strengths of fcc aluminum and copper." *Physical Review Letters* 82: 2713.

Stampfl, C., and C. G. Van de Walle. 1999. "Density-functional calculations for III-V nitrides using the local-density approximation and the generalized gradient approximation." *Physical Review B* 59: 5521.

Umeno, Y., and T. Kitamura. 2002. "Ab initio simulation on ideal shear strength of silicon." *Materials Science and Engineering. B* 88: 79.

Vitos, L., P. A. Korzhavyi, and B. Johansson. 2003. "Stainless steel optimization from quantum mechanical calculations." *Nature Materials* 2: 25.

Wang, J., J. Li, S. Yip, S. Phillpot, and D. Wolf. 1995. "Mechanical instabilities of homogeneous crystals." *Physical Review B* 52: 12627.

Xu, G., A. S. Argon and M. Ortiz. 1995. "Nucleation of dislocations from crack tips under mixed modes of loading: implications for brittle against ductile behaviour of crystals." *Philosophical Magazine A* 72: 415.

Zimmerman, J. A., H. Gao, and F. F. Abraham. 2000. "Generalized stacking fault energies for embedded atom FCC metals." *Modeling and Simulation in Materials Science Engineering* 8: 103.

Further Reading

A similar attempt to connect stacking fault energy with redistribution and topological properties of charge density was made in: Kioussis, N., M. Herbranson, E. Collins, M. E. Eberhart. 2002. "Topology of electronic charge density and energetics of planar faults in fcc metals." *Physical Review Letters* 88, 125501.

8 Fracture Dynamics

Minimum energy paths for unit advancement of a crack front are reaction coordinates in the analysis and understanding of crack-tip mechanics. Results on activation energy barrier and atomic displacement distribution are examined in a comparison of ductile fracture in metal Cu and brittle fracture in semiconductor Si. The contrast illustrates the mechanisms of partial dislocation loop emission versus crack-front extension in a kink-like fashion. The availability of atomistic insights allows a broader appreciation of material complexities of ductile-brittle transitions in solids exposed to extreme environments.

Introduction

Is it possible to probe how a sharp crack in a crystal lattice evolves under stress without actually driving the system to the point of instability? This question is motivated by the basic idea of nondestructive testing. In fracture, this means knowing the pathway of crack-front motion while the lattice resistance against such displacement is still finite. Despite the use of MD simulation to study crack propagation (see, for example, Buehler, Abraham, and Gao 2003), this scenario has not been examined in a broader context. Most studies have been carried out in an essentially 2D setting, with PBC imposed along the direction of the crack front. In such simulations, the crack tip is unphysically constrained due to system size, or equivalently, computational capability limitations. As a consequence, in direct MD simulation of crack initiation and propagation, one is invariably driving the system to instability by applying a loading that is unrealistic. This, in turn, means that the results obtained would be far from the true characteristics of slow crack growth by thermal activation.

Consider an approach capable of following crack-front response without subjecting the system to critical loading. This involves using a reaction pathway sampling method to determine the so-called minimum-energy pathway (MEP), which is the system-level trajectory for the crack front to advance by one atomic lattice spacing, while the imposed load is kept below the critical threshold. In this essay, we discuss the results obtained on crack extension in a metal (Cu) (Zhu, Li, and Yip 2004a) and a semiconductor (Si) (Zhu, Li, and Yip 2004b) to show how ductility or brittleness manifest in the mechanical response of crack front deformation on the nanoscale.

Minimum-Energy Path (MEP) Sampling

Consider a 3D atomically sharp and initially straight crack system subjected to uniaxial tensile load K_I (mode I stress intensity factor) in incremental steps, as depicted in figure 8.1. The crack front would not move spontaneously if the driving force is not sufficient to overcome the intrinsic lattice resistance.

In figure 8.1(b), the open, gray, and closed circles denote the initial state, saddle point, and final state in a barrier activation scenario. With increasing load, the energy landscape begins to tilt in favor of lowering the barrier until one reaches K_{ath} when the barrier has vanished and no activation is then needed for the system to move to the next minimum.

Figure 8.1 shows one can regard the effect of the applied load as tilting the energy landscape toward the final configuration with a corresponding reduction in the activation barrier. So long as the barrier remains finite, the crack front would not move from its initial configuration without assistance from thermal fluctuations. In the simulations discussed below, we are concerned with only situations below K_{ath}, thereby avoiding the problem of over driving.

Figure 8.2 shows the resulting MEP for Cu at a loading of 0.75 K_{at} for two dislocation configurations in the process of dislocation emission from the crack, one corresponding to a straight dislocation (dashed line) and the other for a partial dislocation loop (solid line) as shown in figure 8.3. Notice the saddled point energy is higher in the former which is what one would expect based on line-tension arguments of pushing a straight line compared to a loop.

In mapping the MEP using the Nudged Elastic Band (NEB) method, one obtains the corresponding atomic configurations at every discrete step of the dislocation evolution from the crack front. These are shown in figures 8.3(b) and 8.3(c) below in the dislocation contour development in the shear and opening directions. Such details are insightful for demonstrating how crystalline defect nucleation and propagation can be elucidated at the atomistic level.

In contrast to probing ductile fracture, one may ask what would be the difference with brittle fracture. Corresponding MEP results for a semiconductor crystal, Si, are shown in figure 8.4 (Zhu, Li, and Yip 2004b).

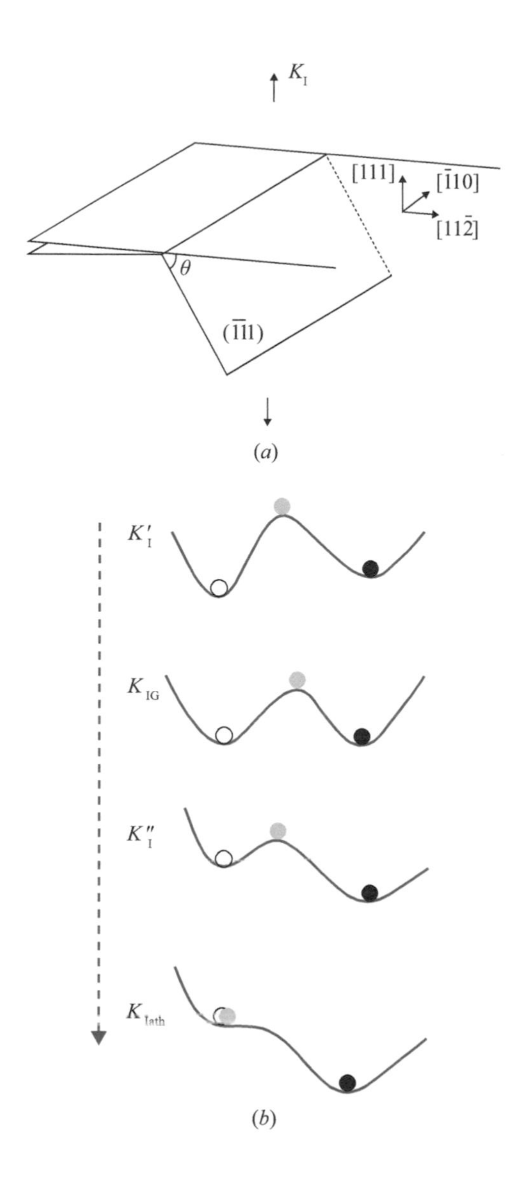

Figure 8.1
(a) Schematic 3D atomically sharp crack front under mode-I (tensile) loading. (b) Energy landscape of the crack system under various loading conditions. $K_{I'}$ is a subcritical load, below the critical value according to the Griffith criterion (Griffith 1921), K_{IG} is the critical load at the saddle point of the landscape, $K_{I''}$ is a supercritical load which is below the athermal threshold value, and K_{Iath} is the *athermal limit* where no thermal activation assistance is needed for the system to go downhill to a nearby minimum. Figure used with permission of ASME, from (Zhu, Li, and Yip 2005); permission conveyed through Copyright Clearance Center, Inc.

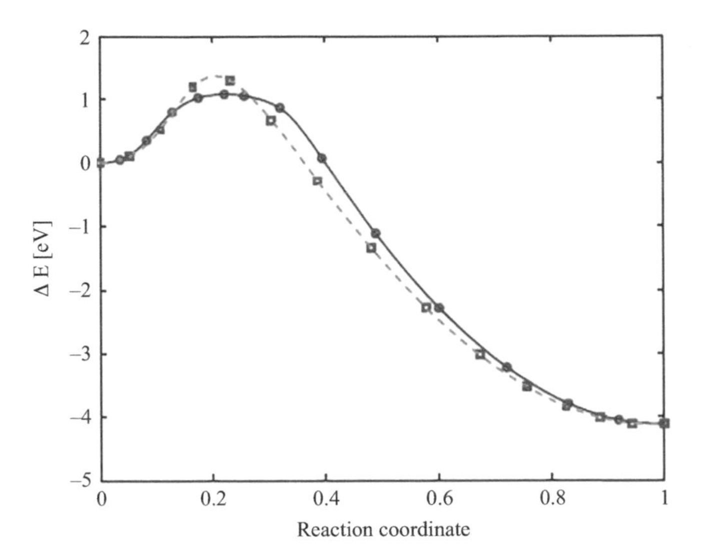

Figure 8.2
MEP for nucleation of a straight dislocation (dashed curve) or a dislocation loop (solid curve) from a crack front in Cu. For further details, see figure 8.3. Reprinted figure with permission from (Zhu, Li, and Yip 2004a). Copyright 2004 by American Physical Society.

The energy landscape is seen to be quite different from that of Cu. Since Si is a brittle solid, a useful reference load is the Griffith value for the stress intensity factor K_{IG} at which the initial and final states are at the same energy (recall figure 8.1b). Note the two states are at the same energy except the crack front has advanced by one atomic spacing in the propagation direction in the final state. By direct calculation, K_{IG} is estimated to be $K_{IG} = 0.646\ \text{MPa}\sqrt{\text{m}}$. This value is lower than the athermal load K_{Iath}, which is due to an effect known as *lattice trapping* (Thomson, Hsieh, and Rana 1971). Being a brittle solid, the relevant deformation mechanism for crack-front advancement is *bond rupture* rather than bond shearing in the case of ductile fracture in Cu. The simulation cell in the Si study contains twenty bonds along the initially straight crack front. It is found the most energetically favored pathway is the breaking of the twenty bonds *sequentially* as seen in figure 8.4 for two different directions of crack propagation. The energy variation is a series of activation barriers each corresponding to the rupture of a bond along the crack front. We see a new feature in the outline of displacement of intermediate magnitude showing a *wedge shape* protruding in the direction of crack-front advancement. The presence of a wedge shape indicates a kink mechanism of crack advancement, namely, the nucleation of a double kink followed by expansion of the left and right kinks in the

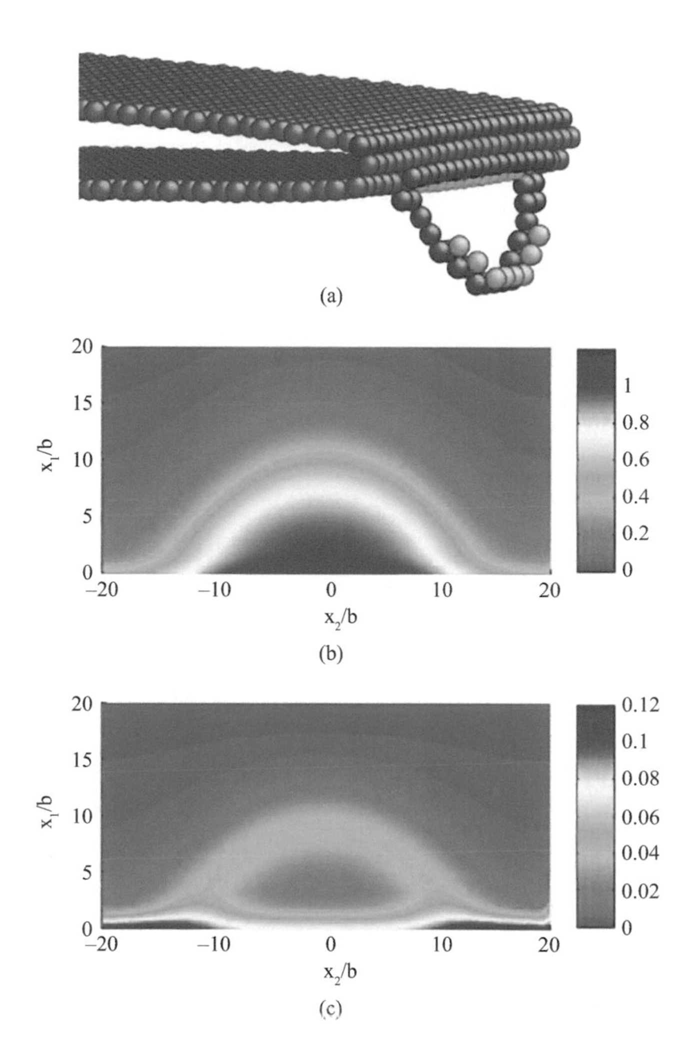

Figure 8.3

(a) Atomistic configuration of an emerging dislocation loop at 75% of the MEP saddle point. Color coding refers to the coordination number N at each atom; light pink $N = 9$, green $N = 10$, dark pink $N = 11$, blue $N = 13$, with perfect coordination $N = 12$ being invisible. (b) Contour plot of shear-displacement distribution normalized by the Burgers vector on the slip plane. (c) Opening-displacement contour normalized by the interplanar spacing. Reprinted figure with permission from (Zhu, Li, and Yip 2004a). Copyright 2004 by the American Physical Society.

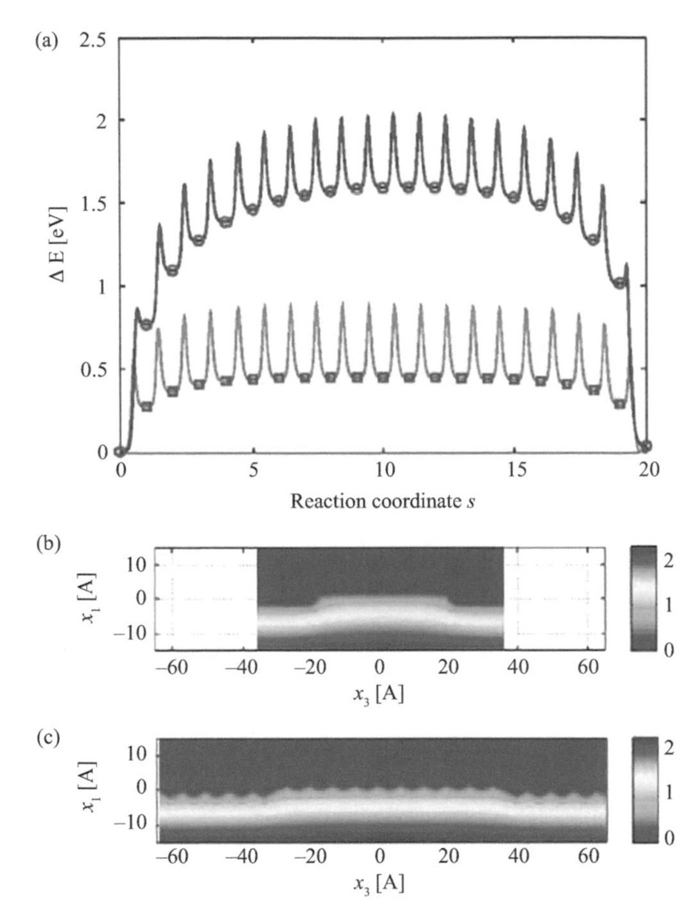

Figure 8.4

MEP and crack-tip geometry for brittle fracture in Si. (a) MEP for the (111)[110] crack (blue upper curve) and the (111)[112] crack (green, lower curve) curve at the respective Griffith loads. (b) Opening-displacement contour across the (111) cleavage plane for the first crack in (a). (c) Same as (b) except for the second crack in (a). Reprinted figure with permission from (Zhu, Li, and Yip 2004b). Copyright 2004 by American Physical Society.

perpendicular direction to the protrusion (recall figures 6.14 and 6.15). It is also significant that this behavior is not seen in figure 8.3 for Cu.

Taking Cu to be a prototypical ductile material, we see that while the crack opening still occurs at the crack front, the large normal displacement lies outside the central region enclosed by the emerging dislocation loop. We attribute this feature to a mechanism of mode switching, or shear-tension coupling, process. The initially large opening displacements in the region swept by the emerging loop are relaxed by giving way to large shear displacements, which are then carried away by the emitted dislocation loop.

It is useful to compare the behavior of atomic displacements at the transition state in Cu and in Si from the standpoint of the nature of interatomic bonding in these two materials. One expects the crack front response in Cu to reflect delocalized metallic bonding, while that in Si should correspond to directional localized covalent bonding. In terms of the energy landscape along the reaction path, one has for Cu a smooth *MEP* with a single major nucleation barrier indicating a crack advancement mechanism as consisting of concerted motions of the atoms to overcome the lattice-resistance barrier by thermal activation. For Si, the existence of significant barriers (cusps in figure 8.4a) is a generic feature of covalently bonded solids. Crack front extension then consists of a series of thermally activated bond ruptures giving rise to kink-pair formation and lateral kink migration along the crack front.

The fact that kink mechanism appears to play a central role in crack front mobility has implications for the analogy between the crack front acting as the core of a sharp crack in a material and the core of a dislocation—both being *line defects* in a crystal lattice. It is well known that dislocation mobility in a directionally bonded solid like Si is governed by thermally activated processes of kink-pair nucleation and migration (Cai et al. 2004) (see essay 6). The present results show a similar mechanism also operates in crack tip advancement, thus reinforcing the notion that line-defect mobility fundamentally depends on crystal structure as well as chemical bonding. From this perspective, the appearance of kink-like structures in figure 8.4(b) is to be expected. Recognition of this underlying connection is significant for the broader appreciation of molecular mechanisms in materials phenomena.

Brittle-Ductile Behavior

In a general-audience discussion of brittle versus ductile fracture, one frequently cannot resist showing dramatic photographs of the hull of a troop transport ship breaking into two when it sailed abruptly into cold waters during World War II. Despite being a well-known phenomenon in the materials community at large, brittle-ductile transition is still an open problem both fundamentally and technologically. Brittle facture was studied early in terms of an estimate of the stress intensity factor which is proportional to the tensile stress loading and the square root of the crack length (Griffith 1921). Ductile fracture, on the other hand, is more complex as it involves the nucleation of a dislocation from the crack front and its subsequent motion away from the crack front (Chiao and Clarke 1989; George and Michot 1993). Criteria for predicting brittle or ductile behavior have been proposed (Rice and Thomson 1974) and later reformulated (Rice 1992) through an interplanar potential for rigid-block sliding in a uniform lattice (Beltz and Rice 1992; Xu, Argon, and Ortiz 1997).

Early MD simulations of brittle-ductile transition have been reported on α-Fe (DeCelis, Argon, and Yip 1983) and later more intensively with improved boundary conditions and detailed characterization of dislocation emission (Cheung and Yip 1990). An essential finding from the later study is shown in figure 8.5. One can readily notice the ductile fracture behavior at the higher temperature and the presence of two dislocations that were emitted from the crack tip. This is an explicit demonstration of the brittle-ductile transition at a temperature lying between 200 and 400 K.

The ductile-brittle transition temperature is a widely known parameter to characterize the brittle-ductile nature of materials. In the case of pure Fe, its value is generally taken to be –50°C or 223 K which is reasonably within the range indicated in figure 8.5. In other words, given the uncertainties associated with classical MD simulation at the time of study, one should not expect any more quantitative determination of materials properties and behavior.

Atomistic simulation allows one to extract detailed information on the energetics and stresses acting on a local inhomogeneity. It has been possible to test the validity of the uniform-displacement assumption in continuum modeling, as well as the assumption of taking the unstable stacking energy γ_{us} (see essay 7) to be the dislocation nucleation barrier (Rice 1992; Cleri et al. 1997). Such an approximation implies that the actual displacement field on the slip plane—nonuniform in all likelihood—may be effectively replaced by a uniform displacement distribution corresponding to rigid-block sliding. Not surprisingly, the assumption of uniform rigid sliding leads to a significantly higher estimate of the energy for dislocation nucleation, a reasonable outcome which, however, could not be quantified without the capability of atomistic simulation regarding removing empiricism (see essay 2). As a result, any assumption pertaining to spatial uniformity and rigid displacement should be suspect. Additionally, simulation has provided insights into crack-tip shielding, the process by which dislocation emission results in local relaxation of the stress field immediately ahead of the crack, thereby increasing the critical load for propagation (Cleri et al. 1997). The implication is that dislocation nucleation and motion is a complex physical mechanism in materials deformation and mechanical response, including crack-tip shielding and blunting, and that atomistic simulation can provide appropriate input to continuum models for predicting brittle versus ductile behavior (Rice 1992; Beltz and Rice 1992; Schoeck 1991; Xu, Argon, and Ortiz 1995; Xu, Argon, and Ortiz 1997).

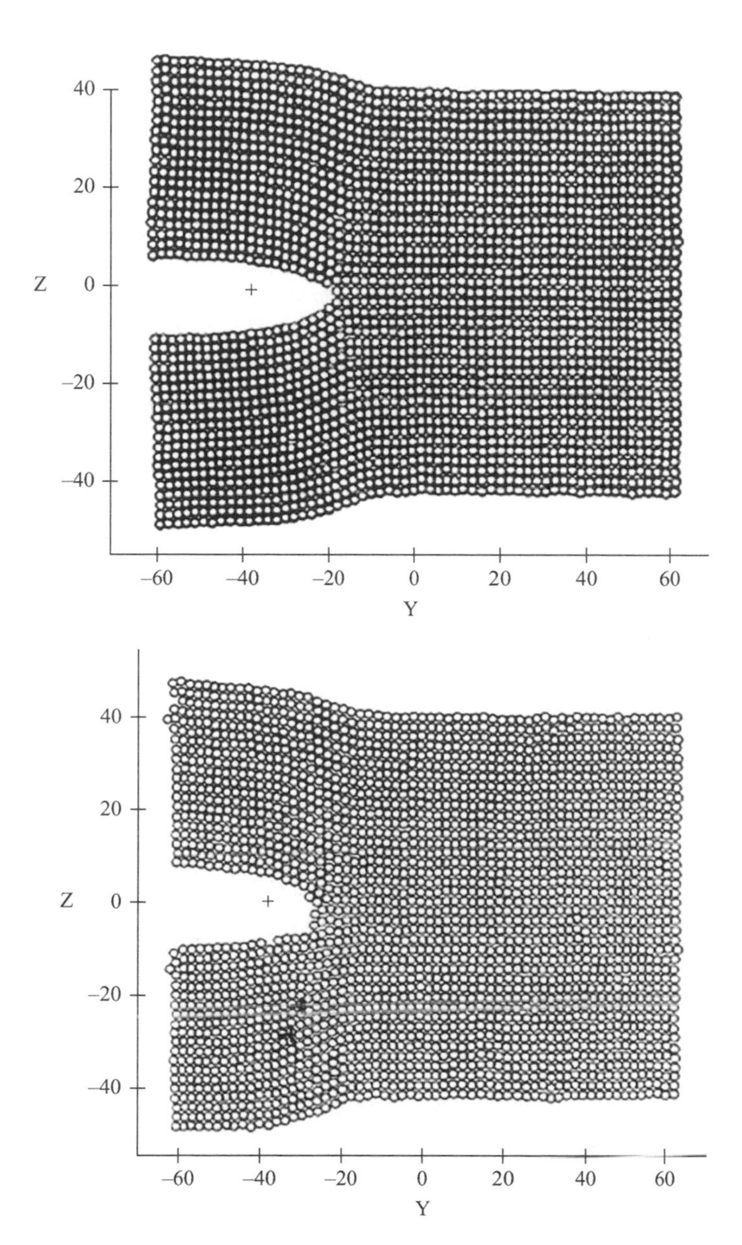

Figure 8.5
Upper figure. Instantaneous atomic configurations of a (100)(011) crack in bcc-Fe at $T = 200$ K and $K_I/K_{IG} = 1.1$, and timestep of 3,000 during MD simulation. The cross denotes the initial position of the crack tip. Lower figure. Same as upper figure except at $T = 400$ K. Two dislocations emitted from the crack tip are indicated. Reprinted figure with permission from (Cheung and Yip 1990). Copyright 2004 by American Physical Society.

References

Beltz, G. E., and J. R. Rice. 1992. "Dislocation nucleation at metal-ceramic interfaces." *Acta Metallurgica et Materialia* 40: S321.

Buehler, M., F. F. Abraham, and H. Gao. 2003. "Hyperelasticity governs dynamic fracture at a critical length scale." *Nature* 426: 141.

Cai, W., V. V. Bulatov, J.-P. Chang, J. Li, and S. Yip. 2004. "Dislocation core effects on mobility." In *Dislocations in Solids*, edited by F. R. N. Nabarro and J. P. Hirth, Vol. 12, chap. 64, 1–80. Amsterdam: Elsevier.

Cheung, K. S., and S. Yip. 1990. "Brittle-ductile transition in intrinsic fracture behavior of crystals." *Physical Review Letters* 65: 2804.

Chiao, Y. H., and D. R. Clarke. 1989. "Direct observation of dislocation emission from crack tips in silicon at high temperatures." *Acta Metallurgica* 37: 203.

Cleri, F., S. Yip, D. Wolf, and S. Phillpot. 1997. "Atomic-scale mechanism of crack-tip plasticity: dislocation nucleation and crack-tip shielding." *Physical Review Letters* 79: 1309.

DeCelis, B., A. S. Argon, and S. Yip. 1983. "Molecular dynamics simulation of crack tip processes in alpha-iron and copper." *Journal of Applied Physics* 54: 4864.

George, A., and G. Michot. 1993. "Dislocation loops at crack tips: nucleation and growth—an experimental study in silicon." *Materials Science and Engineering. A* 164: 118.

Griffith, A. A. 1921. "The phenomena of rupture and flow in solids." *Philosophical Transactions of the Royal Society of London* A 221: 163–198.

Rice, J. R. 1992. "Dislocation nucleation from a crack tip: an analysis based on the Peierls concept." *Journal of the Mechanics and Physics of Solids* 40: 239.

Rice, J. R., and R. Thomson. 1974. "Ductile versus brittle behaviour of crystals." *The Philosophical Magazine: A Journal of Theoretical Experimental and Applied Physics* 29: 73.

Schoeck, G. 1991. "Dislocation emission from crack tips." *Philosophical Magazine A* 63: 111.

Thomson, R., C. Hsieh, and V. Rana. 1971. "Lattice trapping of fracture cracks." *Journal of Applied Physics* 42: 3145.

Xu, G., A. S. Argon, and M. Ortiz. 1995. "Nucleation of dislocations from crack tips under mixed modes of loading: implications for brittle against ductile behaviour of crystals." *Philosophical Magazine A* 72: 415.

Xu, G., A. S. Argon, and M. Ortiz. 1997. "Critical configurations for dislocation nucleation from crack tips." *Philosophical Magazine A* 75: 341.

Zhu, T., J. Li, and S. Yip. 2004a. "Atomistic study of dislocation loop emission from a crack tip." *Physical Review Letters* 93: 025503.

Zhu, T., J. Li, and S. Yip. 2004b. "Atomistic configurations and energetics of crack extension in silicon." *Physical Review Letters* 93: 205504.

Zhu, T., J. Li, and S. Yip. 2005. "Nanomechanics of crack front mobility." *Journal of Applied Mechanics* 72: 932.

Further Reading

Cheung, K. S., and S. Yip. 1994. "Molecular dynamics simulation of crack-tip extension: brittle-ductile transition." *Modelling and Simulation of Materials Science Engineering* 2: 865.

Tang, M., and S. Yip. 1994. "Lattice instability in SiC and simulation of brittle fracture." *Journal of Applied Physics* 76: 2719.

Tang, M., and S. Yip. 1995. "Atomic size effects in pressure-induced amorphization of SiC." *Physical Review Letters* 75: 2538.

Belak, J. 2002. "Multi-scale applications to high strain-rate dynamic fracture." *Journal of Computer-Aided Materials Design* 9: 165–172. Ductile dynamic fracture, multiscale modeling, void growth, spallation.

Romano, A., J. Li, and S. Yip. 2006. "Atomistic simulation of rapid compression of fractured SiC." *Journal of Nuclear Materials* 352: 22.

Silva, E. C. C. M., J. Li, D. Liao, S. Subramanian, T. Zhu, and S. Yip. 2006. "Atomic scale chemo-mechanics of silica: nanorod deformation and water reaction." *Journal of Computer-Aided Materials Design* 13: 135–159. Transition state pathway sampling and molecular orbital theory study of hydrolytic attack of an SiO_2 nanorod focused on bond rupture and the role of bond strain in initiation.

Silva, E., C. Först, J. Li, X. Lin, T. Zhu, and S. Yip. 2007. "Multiscale materials modelling: Case studies at the atomistic and electronic structure levels." *ESAIM: Mathematical Modelling and Numerical Analysis* 41 (2): 427–445. Fundamental problems to illustrate scientific challenge with technological relevance involving two current themes, mining of atomistic simulations to extract insight into mechanical failure of materials and manipulation of electronic structure effects for nanomaterial innovation.

9 Interface Strength

While experimental data already suggested the traditional Hall–Petch scaling with grain size may not hold in nanocrystalline metals, it remains for atomistic simulation to show in a quantitative fashion that a reverse effect occurs when the grain size is in the range of ten nanometers. It is physically intuitive that the transition signals an intragranular to intergranular crossover. Details from MD simulation provide a quantitative explanation of the cross over behavior with implications for the role of grain-boundary sliding and migration as mechanisms controlling interfacial strength. Additionally, the existence of a critical grain size is analogous to the ductile-brittle transition temperature in fracture discussed in the preceding essay.

Introduction

The strength of polycrystalline materials can vary significantly with the size of the crystalline grains. In materials with grain size of micrometers or larger, deformation is known to proceed through intragranular mechanisms of dislocation reactions and pile up at the GBs. The result is an increasing strength with decreasing grain size, a behavior known as the Hall–Petch effect. However, at the nanometer scale, intergranular processes become dominant, then strength decreases with decreasing grain size. Understanding and predicting the grain size where this reversal manifests has been of interest to the broad community of materials science and technology, in a way similar to the determination of ductile-brittle transition temperature discussed in essay 8.

Nanocrystal Deformation—Reverse Hall–Petch Behavior

Interests in nanoscience and technology coupled with multiscale modeling and simulation in the materials community (computational materials) have had significant impact on our understanding of deformation processes operating at material interfaces. A major opportunity is the optimization of strength and ductility (Siegel 1994a). Typically, hardness and ductility increase with decreasing grain size. On the other hand, experimental

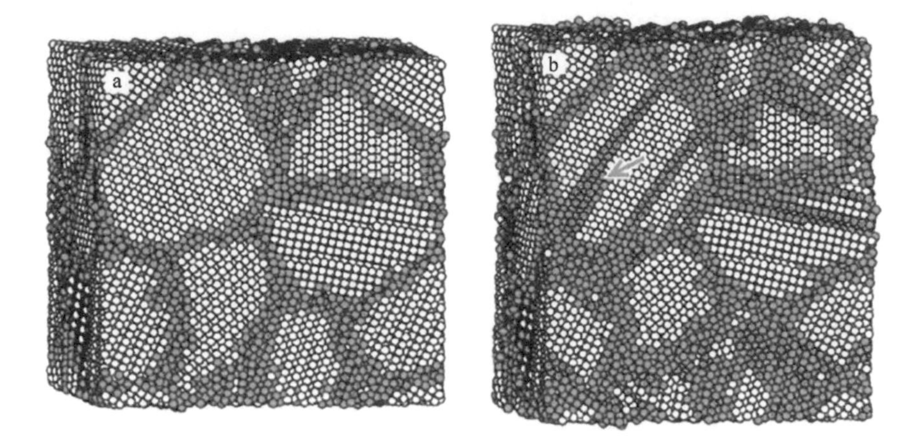

Figure 9.1
Nanocrystalline sample in MD simulation before (left) and after (right) deformation at 10% strain. Color coding shows GB thickening (blue) and intragranular dislocations slips (red). Reprinted by permission from Springer Nature: (Schiotz, Di Tolla, and Jacobsen 1998).

evidence also exists showing the reverse effect could occur when grain sizes are in the few-nanometer range (Chokshi et al. 1989; Siegel 1994b). Around 1998, there was no general consensus on what would be the mechanism for such an effect. MD simulation indicated strength softening with grain size was possible in the presence of many small sliding events at the interfaces, as can be seen in figure 9.1 before and after a 10% tensile deformation (Schiotz, Di Tolla, and Jacobsen 1998). Detailed analysis showed the softening was associated with a reverse Hall–Petch scaling behavior with grain size, figure 9.2.

The Strongest Size

From the perspective of an intragranular to intergranular transition, the crossover to reverse Hall–Petch implies the existence of a "strongest size," an optimization concept in nanoscience and technology illustrated in figure 9.3, along with experimental results on Cu pointing to this phenomenon in figure 9.4.

Mechanism Map

Additional simulation studies of the deformation mechanisms revealed a complex dependence on the stacking fault energy (recall essay 7), the elastic properties of the metal, and the magnitude of the applied stress. These considerations led to the construction of

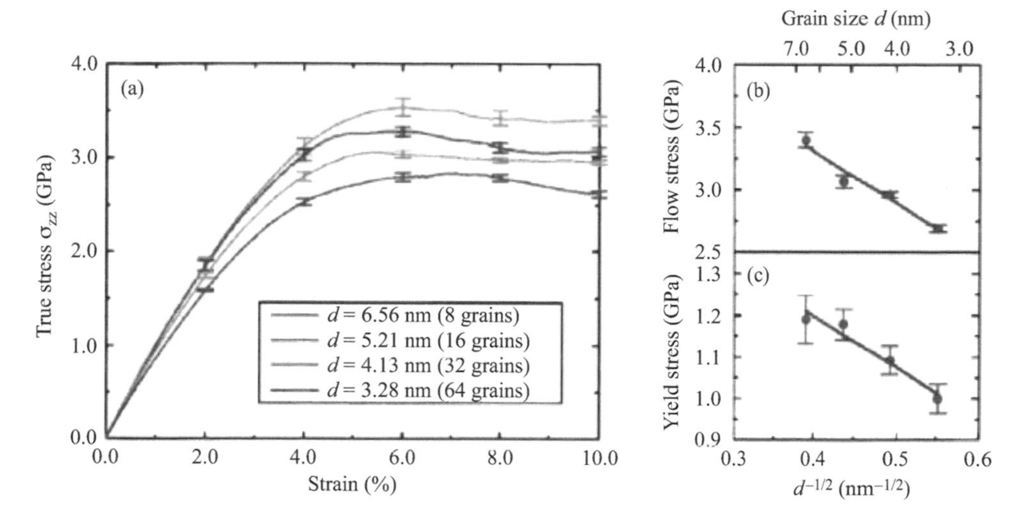

Figure 9.2

(a) Stress-strain behavior at four grain sizes showing grain-size softening. Flow (b) and yield (c) stress scaling with grain size showing a reverse Hall–Petch effect. Reprinted by permission from Springer Nature: (Schiotz, Di Tolla, and Jacobsen 1998).

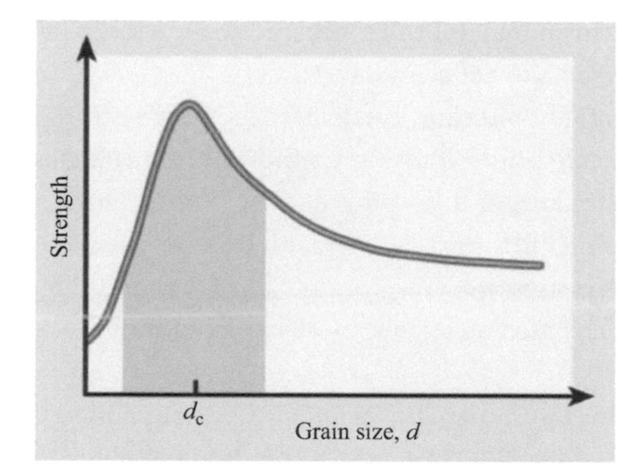

Figure 9.3

Schematic variation of the expected strength of a nanocrystal with grain size in the few nanometer range. With decreasing size, the strength first increases, which is the normal Hall–Petch effect, and then it reaches a maximum before decreasing quite sharply, which may be regarded as a reverse Hall–Petch effect. This crossover essentially signifies the transition from an *intragranular strengthening* mechanism to an *intergranular softening* mechanism. Figure from (Yip 1998).

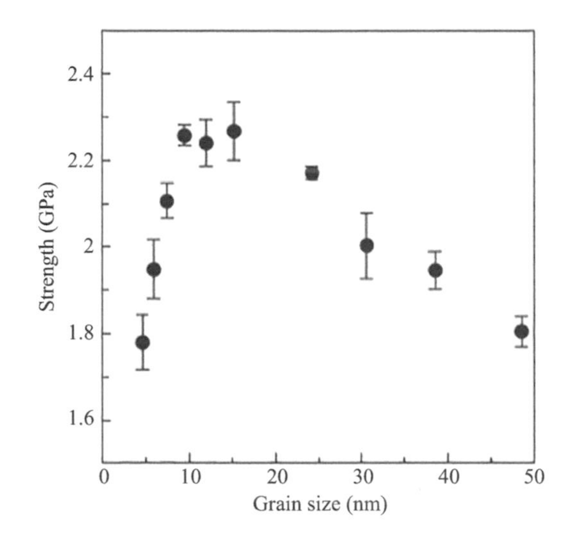

Figure 9.4
Grain size scaling of flow stress in *experiment* showing the transition from Hall–Petch to reverse Hall–Petch behavior and the existence of "the strongest size" in Cu (Chokshi et al. 1989; Yip 2004). Compare these data with the schematic variation in figure 9.3. Figure from (Yip 2004).

a three-region mechanism map for the mechanical behavior of nanocrystalline metals, shown in figure 9.5 (Yamakov et al. 2004).

The 1/d line essentially indicates the existence of "the strongest size," playing the analogous role of ductile-brittle transition temperature in fracture (essay 8). Basically, figure 9.5 shows dominance of dislocation process at large grain size and a distinction between metals with high intrinsic stacking fault energies such as Al and those with low γ_{SF}. At decreasing grain sizes, the GB processes of sliding and migration become dominant across the 1/d line based on an empirical scaling of the nucleation stress with grain size (Yamakov et al. 2004).

How crystal grains deform in a material under stress is important both scientifically, for the understanding of plastic flow in solids, and technologically, for controlling materials strength and ductility. In this context, grain size plays an essential role. In polycrystals with grain size in the micrometer range, strength increases with decreasing size. The effect is attributed to the pileup of dislocations at the GBs. With the synthesis of nanocrystals, now with grains in the nanometer range, the opposite behavior was found experimentally first and then confirmed by atomistic simulation. This is a demonstration of how molecular details provided by simulation can provide atomic-level insights into materials phenomena.

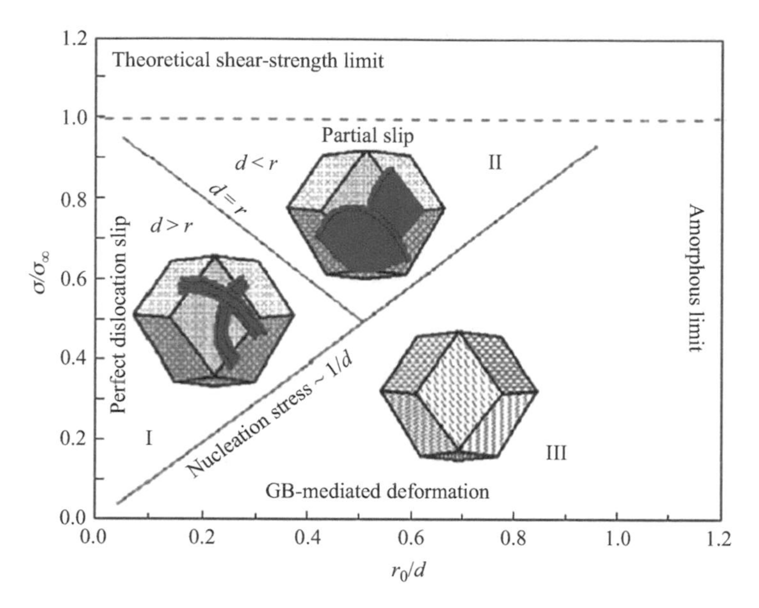

Figure 9.5
Mechanism map for tensile deformation of nanocrystalline fcc metals showing three regimes: full (I) and partial (II) dislocation dominant, and GB sliding and migration dominant (III). Reprinted by permission from Springer Nature: (Yamakov et al. 2004).

It is not surprising that the variation of strength or hardness with grain size should have a characteristic length scale. Under large strain deformation, intragranular and intergranular mechanisms can be activated and, in general, will compete with each other. As each has a different dependence on grain size, there will be a transition region when one takes over from the other. The behavior is considered "normal" when the grain size d is greater than the characteristic length d_c (figure 9.3). The reverse effect, when $d < d_c$, simply signifies a change in the dominant deformation mechanism. When such crossover conditions arise, one can speak of microstructural heterogeneities as a degree of materials complexities (absent in single crystals).

The behavior of granular materials near d_c can be particularly interesting when functional properties are coupled to the microstructural heterogeneities, such as epitaxial films and quantum dots (Schowalter 1996). In these heterostructures, the lattice mismatch between different materials will induce strain, and how it relaxes during fabrication will determine the material structure and its functionalities. Again, there is competition and tradeoff between processes mediated by misfit defects such as dislocations, and surface-based processes like roughening and morphological modifications.

As a final perspective and again recalling the comments made in essay 2 on what makes atomistic simulations unique, we see that simulations can provide insightful details at the atomic level not accessible by other means. Advances in both experimental techniques and computational capabilities will continue to close the gap between measurements and simulation. What stands to benefit from either side is our physical understanding and ability to innovate and optimize the materials needed by society.

References

Chokshi, A. H., A. Rosen, J. Karch, and H. Gleiter. 1989. "On the validity of the Hall-Petch relationship in nanocrystalline materials." *Scripta Metallurgica* 23: 1679–1684 (1989).

Schiotz, J., F. D. Di Tolla, and K. W. Jacobsen. 1998. "Softening of nanocrystalline metals at very small grain sizes." *Nature* 391: 561–563.

Schowalter, L. J., ed. 1996. *Heteroepitaxy and Strain. MRS Bulletin (Special Issue)* 21 (4).

Siegel, R. W. 1994a. "Nanophase Materials." In *Encyclopedia of Applied Physics* Vol. 11, 173–199. New York: VCH.

Siegel, R. W. 1994b. "What do we really know about the atomic-scale structures of nanophase materials?" *Journal of Physics and Chemistry Solids* 55: 1097–1106.

Yamakov, V., D. Wolf, S. R. Phillpot, A. K. Mukherjee, and H. Gleiter. 2004. "Deformation-mechanism map for nanocrystalline metals by molecular-dynamics simulation." *Nature Materials* 3: 43–47.

Yip, S. 1998. "The strongest size." *Nature* 391: 532–533.

Yip, S. 2004. "Mapping plasticity." *Nature Materials* 3: 11–12.

Further Reading

Hall, E. O. 1951. "The deformation and ageing of mild steel: III discussion of results." *Physics Society of London B* 64: 747–753. Landmark paper on the Hall–Petch effect.

Petch, N. J. 1953. "The cleavage of polycrystals." *Journal of the Iron and Steel Institute* 174: 25–28. Landmark paper on the Hall–Petch effect.

Yip, S., and D. Wolf. 1989. "Atomistic concepts for simulation of grain boundary fracture." *Materials Science Forum* 46: 77. A strategy to exploit four related methods of materials analysis, lattice statics to determine GB energies, lattice dynamics to analyze local elastic constants, MC simulation to determine solute segregation in GB structures, and MD simulation to model dynamic crack extension.

Wolf, D., and S. Yip. 1990. "Interfaces part I: structure, chemistry, electronic properties." *Materials Research Society Bulletin* 15: 21. Theme issue in two parts.

Wolf, D., and S. Yip. 1990. "Interfaces part II: mechanical and high-temperature behavior." *Materials Research Society Bulletin* 15: 38.

Nieh, T. G., and J. Wadsworth. 1991. "Hall-Petch relation in nanocrystalline solids." *Scripta Metal Materials* 25: 955–958. Early commentary on the inverse Hall–Petch behavior.

Nguyen, T., P.S. Ho, T. Kwok, C. Nitta, and S. Yip. 1992. "Thermal structural disorder and melting at a crystalline interface." *Physical Review B* 46: 6050.

Wolf, D., S. Phillpot, J. M. Rickman, and S. Yip. 1992. "On the relation between sliding and migration for high-angle grain boundaries." *Materials Science Forum* 94–96: 487.

Wolf, D., and S. Yip, eds. 1992. *Materials Interfaces—Atomic-level Structure and Properties*. Cambridge, UK: Chapman & Hall. A compendium of twenty-seven chapters by forty-seven contributors in four parts: Bulk Interfaces, Semi-Bulk and Thin Film Interfaces, Role of Interface Chemistry, and Fracture Behavior.

Yip, S. 1992. "Simulation studies of interfacial phenomena—melting, stress relaxation and fracture." In *Molecular Dynamics Simulations*, edited by F. Yonazawa, 221. Berlin: Springer-Verlag.

Shen, Y. F., L. Lu, Q. H. Lu, Z. H. Jin, and K. Lu. 2005. "Tensile properties of copper with nanoscale twins." *Scripta Materialia* 52: 989. Strength and ductility increase with decreasing twin lamella thickness, following Hall–Petch relationship.

Argon, A. S., and S. Yip. 2006. "The strongest size." *Philosophical Magazine Letters* 86: 713. Mechanism modeling analysis.

Chen, J., L. Lu, and K. Lu. 2006. "Hardness and strain-rate sensitivity of nanocrystalline Cu." *Scripta Materialia* 54: 1913.

Yildiz, B., A. Nikiforova, and S. Yip. 2009. "Metallic interfaces in harsh chemo-mechanical environments." *Nucleur Engineering Technology* 41: 21.

Albe, K., Y. Ritter, and D. Sopu. 2013. "Enhancing the plasticity of metallic glasses: Shear band formation, nanocomposites and nanoglasses investigated by molecular dynamics simulations." *Mechanical Materials* 67: 94. Tuning mechanical properties by solute, precipitate, and GB insertions reveals glass-glass interfaces act as structural heterogeneities to promote shear-band formation and prevent strain localization.

Armstrong, R. W. 2014. "60 years of Hall-Petch: past to present nano-scale connections." *Materials Transactions* 55: 2. Overview of developments leading to current improvements in strength of metals with nanoscale grains.

Palkovic, S. D., S. Moeini, S. Yip, and O. Büyüköztürk. 2015. "Mechanical behavior of a composite interface: calcium silicate hydrates." *Journal of Applied Physics* 118: 034305.

Ioannidou, K., K. J. Krakowiak, M. Bauchy, C. G. Hoover, E. Masoero, S. Yip, F.-J. Ulm, P. Levitz, R. J.-M. Pellenq, and E. Del Gado. 2016. "Mesoscale texture of cement hydrates." *Proceedings of the National Academy of Science* 113: 2029.

IV Viscous Relaxation

10 Metadynamics Simulation: Viscous Liquids

Extending the timescales associated with traditional MD simulation is a continuing challenge to the community of multiscale materials modeling and simulation. Many important properties and behavior of materials in global science and technology need to be understood and manipulated over spatial and temporal scales, ranging from the electronic structure and the molecular (microscale), the finite elements (mesoscale), and the continuum (macroscale). Given the timescales of MD simulation fall in the microscale range, acceleration methods had been developed to extend into the mesoscale and beyond. The focus of this essay is an alternative approach, based on the concept of metadynamics. The basic idea is to estimate (sample) the probability that the model system will activate over potential energy barriers. Instead of moving particles individually according to the Newtonian forces exerted on them by their neighbors, the system now evolves on a potential energy surface (PES) (landscape) without having to compute explicitly the interatomic forces for the particles. This metadynamics approach is stochastic in nature, and the resulting algorithm allows molecular simulations to be carried out over timescales many orders of magnitude longer than the conventional MD method.

Introduction

One of the most vexing challenges in MD simulations is the time-scale bottleneck. While mechanisms operate at all length timescales, the most critical information in our study of molecular mechanisms often lies in those interaction details at the molecular level, which in turn will trigger subsequent behavior that manifest at much longer timescales. By long timescales, we have in mind distinguishing several time regions: subpicoseconds to nanoseconds to microseconds constitute the microscale range, milliseconds to minutes and hours delineate the mesoscale, and days and beyond the macroscale, respectively.

The limitation of MD stems from the time-step size chosen in the integration of Newton's equations of motion. Typically, time steps of femtoseconds are used, which makes it prohibitive for small research groups to computationally follow the system evolution longer than tens or hundreds of nanoseconds. Exceptions could be the scientists at national laboratories with access to dedicated high-performance computing facilities. This means that explicit molecular simulations of mesoscale behavior are not readily feasible in practice.

We will focus on the implementation of metadynamics in specific case studies of molecular processes occurring spatially on the nanoscale and temporally on the meso- and macroscales. Here and in essay 11, we will address the well-known problem of the shear viscosity of liquids undergoing vitrification (the glass transition). Essay 12 will treat rate-dependent nucleation of defects in crystalline solids. Essays 13–15 will examine the molecular mechanisms in the rheological behavior of noncrystalline solids. All these studies are based on the long time-scale simulation methodology discussed in the present essay.

We will describe the metadynamics simulation method in the specific case study of the viscosity of highly viscous liquids, a well-known challenge in the community of statistical physics and theoretical chemistry. A basin-filling algorithm tracks the system as it climbs out of one local, deep-energy minima after another through a series of activation and relaxation steps, the process repeating autonomously over the entire simulation run. The algorithm generates a system trajectory on the PES, which we will denote as the transition state pathway (TSP). In other words, the system evolution "data" consist of a sequence of energies and a corresponding set of particle coordinates. Any system property or behavior of interest can be determined by postprocessing the TSP data.

In using the TSP data, we consider two approaches differing in theoretical rigor. The first is a heuristic mean-field description that yields a coarse-grained temperature-dependent *activation barrier*. This approach is able to account qualitatively for the so-called *fragile* behavior commonly regarded as the dynamical signature of the liquid-to-glass transition. The second approach is more theoretically sound but technically more challenging. It involves formulating a Markov network model. It is capable of giving viscosity results for direct comparison with simulation findings in the low-viscosity regime, as well as with experimental measurements in the high-viscosity regime (10^2–10^{12} $Pa \cdot s$).

From the standpoint of molecular understanding of transport phenomena, the present essay demonstrates what we mean by extending the reach of traditional MD simulations. In the following companion essay, we consider further the behavior of viscous silica, a representative *strong* glass forming liquid. Combining the two essays—explaining both fragile and strong behavior self-consistently—provides a unifying mechanistic perspective on the nature of the glass transition.

The viscosity of a liquid, being a product of the shear modulus and a structural relaxations time, is a measure of the shear stress relaxation in the system. When viscosity increases strongly with lowering temperature, the behavior is attributed mostly to a rapidly increasing relaxation time, a manifestation of the onset of slow dynamics of viscous

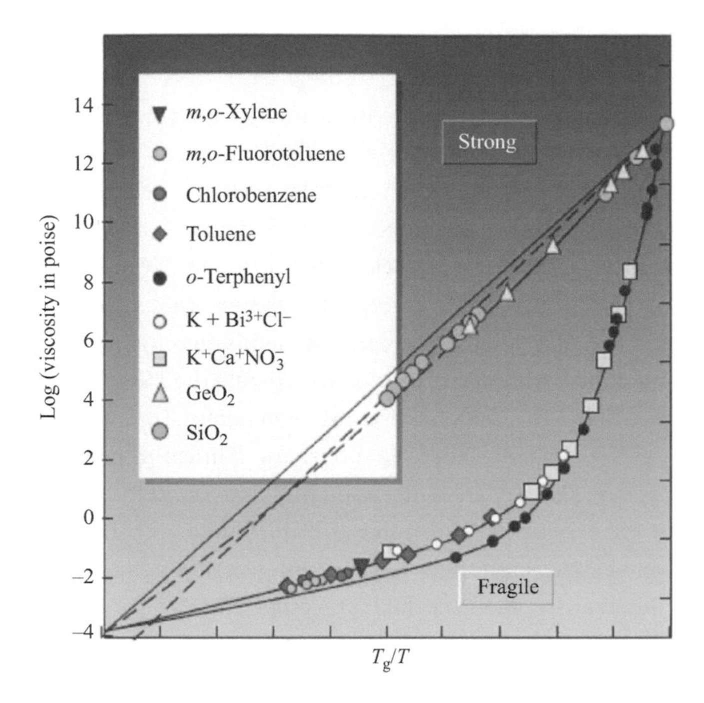

Figure 10.1
Strong and fragile behavior of selected supercooled liquids. Reprinted by permission from Springer Nature: (Debenedetti and Stillinger 2001).

flow. Two questions have been debated in the community concerning this ubiquitous phenomenon. One is *conceptual*—a slowing down in dynamics is a signature of the *glass transition*. More will be said about this below and also in essay 11. The other challenge is *computational*—how can one predict the viscous behavior from a knowledge of molecular structure and intermolecular interactions. This is the challenge we will address here.

To provide context for the viscous behavior of liquids quenched below their freezing point (super-cooled) we begin with the well-known body of experimental results on two types of glass-forming liquids, shown in figure 10.1. Notice the sharp contrast in the temperature dependence between the two groups, one following an Arrhenius behavior (will be referred to as *strong*) while the other showing a characteristically super-Arrhenius (fragile) variation. A simple physical question, suitable for students and experts alike, is what is the reason for this stark difference?

Many, although not all liquids, upon undercooling, exhibit a super-Arrhenius shear viscosity behavior over a narrow temperature range, a characteristic known as *fragility* (Angell 1995). The fragile behavior is of interest to molecular-level calculations that sample the underlying PES (Grigera et al. 2002; Coslovich and Pastore 2007). While the minima of the PES (inherent structure) (Stillinger and Weber 1982) and saddle points can be found by various numerical procedures, no method yet exists that can routinely compute the viscosity behavior over the viscosity range from 10^2 to 10^{12} Pa · s.

A sampling algorithm that allows a molecular system to climb out of an arbitrary series of potential wells following a minimum energy path was first proposed in 2002 (Laio and Parrinello 2002; Martoňák, Laio, and Parrinello 2003). In this method, one applies energy penalty functions to activate the system to continuously explore new regions of phase space. As a variant, one can couple this algorithm of activation through energy penalty function with static relaxation to generate the TSPs of super-cooled liquids (Kushima et al. 2009a). This is the version considered herein. For a systematic application of the TSP trajectory we adopt a heuristic model and a linear response theory approach (McQuarrie 1973) based on a Markov network. Both are found to be capable of accounting satisfactorily the experimental features of *fragile* liquids. In the following essay, we undertake a study of silica, known for its representative *strong* liquid behavior in viscosity, using only the heuristic method. Taken together, the two studies demonstrate how the difference between *strong* and *fragile* glass formers can be traced to the roughness of the PES, thus providing quantitative computational support for the fundamental nature of the landscape perspective (Stillinger 1995; Sastry, Debenedetti, and Stillinger 1998).

Basin-Filling Algorithm—Autonomous Basin Climbing (ABC)

Figure 10.2 is a schematic illustration of the essence of meta-dynamics simulation, a method capable of lifting a given system out of any potential well via a series of activation-relaxation steps. The algorithm is a modification of that originally introduced to escape from free-energy minima (Laio and Parrinello 2002); the Laio-Parrinello concept has come to be known as *metadynamics* (Martoňák, Laio, and Parrinello 2003).

Consider starting the system in a particular local energy minimum E^m with corresponding atomic configuration $\underline{r}^m$. From this initial state, an activation step is applied to drive the system away from its current configuration by imposing a prescribed energy penalty function (specified below), followed by a relaxation step to allow the system to settle into an energy-minimized configuration in the presence of the penalty function. At the end of each activation-relaxation sequence, the system will find itself in a different energy state, typically higher than before the penalty imposition, with a new set of relaxed atomic configurations. This process of activation-relaxation is repeated until the system finds itself in an appreciably lower energy state, which after checking (see below), indicates a new local minimum nearby to E^m has been reached. With this new local minimum and the corresponding system configuration one can backtrack to deduce the

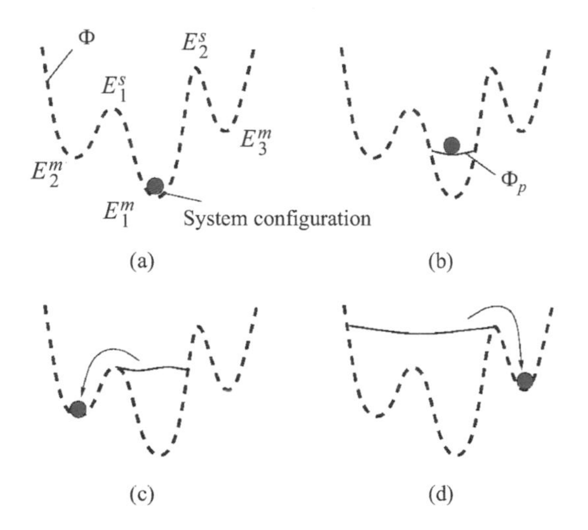

Figure 10.2
Schematic illustration of the essence of meta-dynamics simulation the basin-filling method. Dashed and solid lines indicate original PES and penalty potential, respectively. Penalty functions push the system out of a local minimum to a neighboring minimum by crossing the lowest saddle barrier (Liao and Parrinello 2002; Kushima et al. 2009a). Reprinted from (Kushima et al. 2009a), with the permission of AIP Publishing.

saddle point E^s and its configuration. The sequence of starting from an initial local minimum E_i^m to crossing a saddle point E_i^s to reaching a nearby local minimum E_{i+1}^m can be repeated to generate a trajectory $\Gamma(E^m, E^s) \equiv (E_1^m - E_1^s - E_2^m - E_2^s - \ldots)$. Figure 10.2 depicts such a process for sampling three local minima and two saddle points.

The algorithm for the basin-filling method can be readily coded:

(a) Select an initial local minimum E_1^m (could choose an inherent structure energy from $\{\alpha\}$) and corresponding system configuration, say $\underline{r}_1^m$.

(b) Activation step—apply a penalty function $\phi_1(\underline{r}) = W \exp\left[\frac{(\underline{r} - \underline{r}_1^o)^2}{2\sigma^2}\right]$

where $\underline{r}_1^o = \underline{r}_1^m$, such that total system energy becomes $\Phi_p^1 = \Phi + \phi_1$. W and σ are prescribed constants which determine the strength and spatial extent of the activation, respectively.

(c) Relaxation step—minimize Φ_p^1 to obtain new energy and configuration $\underline{r}_2^o$

(d) Repeat (b) and (c) with $\Phi_p^2 = \Phi_p^1 + \phi_2$, $\Phi_p^3 = \Phi_2 + \phi_3$, ... until the difference between Φ_p and Φ vanishes.

(e) Confirm the sampling of a new local minimum E_2^m and configuration $\underline{r}_2^m$ by checking to see both $\frac{\partial \Phi}{\partial r}(\underline{r}_2^m) = 0$ and $\phi_p(\underline{r}_2^m) = 0$ are satisfied.

Notice that when the penalty functions are added to the system, as long as the system has not climbed out of the well, there will be a difference between Φ_p and Φ. When the system does escape from the well and goes into a new minimum, it may not be clear by looking at Φ_p and Φ alone, but their difference should be essentially zero. One needs to check after every relaxation to see if the two conditions listed in (e) are satisfied. If they are both satisfied, this means the system is indeed in a new minimum. If not, that means the system has not escaped. The two conditions are necessary and sufficient, while the vanishing of the difference between Φ_p and Φ is necessary but not sufficient.

In the present algorithm we do not specify the search direction as in the dimer method (Henkelman and Jonsson 1999). By keeping all the Gaussian penalty functions imposed during the simulation, we eliminate frequent recrossing of small barriers, which is a significant advantage of the history-penalized basin-filling methods (Laio and Parrinello 2002; Wang and Landau 2001). On the other hand, this type of method could suffer from poor scaling with system size or number of particles participating in the activation-relaxation. In principle, it may seem that the number of penalty functions needed to perform adequate sampling could grow exponentially with the configuration space dimension.

We have found this is not the case in the filling of glassy basins because the volume of configuration space to be sampled is much smaller than that of the entire system. The reduction arises from the particles being effectively constrained to their local atomic positions. For a crystal where each atom occupies $(x\%)^3$ of the configuration space, the volume to be sampled is $(x\%)^{3N}$. From the Lindemann criterion for melting (Löwen 1994), an estimate of $x\%$ is approximately 0.13 for face-center-cubic crystals. For the binary Lennard-Jones liquid, the range of atomic mobility may be obtained from the mean-square displacement determined by MD simulation (Kob and Andersen 1995). The mean-square displacement shows quadratic variation at short times corresponding to ballistic motion of particles. At intermediate times, the mean-square displacement levels off to a plateau because the particle is trapped in the cage formed by the surrounding neighboring particles. Once the system gets out from this cage, the mean-square displacement will start to increase again. One can estimate the size of the cage to be $\langle r^2(t)\rangle \sim 0.1$. We therefore chose $\sigma^2 = 0.1$ corresponding to the estimated cage size. Using penalty functions with σ estimated in this manner we found that about 2,000 penalty functions are sufficient to fill the lowest minimum found in the inherent structure analysis using simulation cells containing 100 and 256 particles. The local minimum well is known to have a volume that depends on the energy, the volume becoming exponentially large as the energy is lowered (Massen and Doye 2007). Therefore, σ needs to be modified according to the energy region of PES sampled (smaller σ for high Φ) in order to escape a single local minimum. On the other hand, we are not interested in activation process for escaping from each local minimum but in escaping from a broad collection of several minima (the basin or cage). Experience shows a compromise is to choose a fixed value of σ to be the estimated size of the cage.

We should emphasize in the present method that the system moves in the energy space Φ_p, which includes all the penalty functions that have been applied up to that point. Since the corresponding system configurations at any stage of the activation and relaxation step is known, we can always display the trajectory track in Φ-space. The distinction between Φ and Φ_p is essential to understanding the results of our algorithm. With the energy landscape sampling proceeding under the dynamics governed by Φ_p, we may consider our method to be based on so-called metadynamics (Laio and Parrinello 2002). Moreover, the sampled trajectory consisting of local minima and saddle points will be displayed only in Φ-space, so it is meaningful to compare our results with those obtained using other methods such as hyperdynamics or adaptive kinetic MC.

A typical trajectory generated by the algorithm just described, applied to a system of 100 particles interacting through the binary Lennard-Jones potential (Kob and Andersen 1995) with PBC, is shown in figure 10.3, using a value of $W = 1.0$ and $\sigma^2 = 0.1$ in reduced units. The sampling started in a low-energy minimum denoted by ⓐ. The energy and corresponding atomic configurations at this state were obtained from an inherent structure calculation at $T = 0.5$. One sees a number of general features of this PES trajectory, such as the irregular and appreciable fluctuations between local minima and saddle points. This particular illustration shows two deep minima (basins), marked as ⓐ and ⓑ, which are well separated along the trajectory. The overall appearance of the trajectory is one of roughness.

Our interest lies in quantifying the connectivity of the local minima through the saddle points. This information is embedded in any trajectory describing the hopping from one major basin to another.

Transition State Pathway (TSP)

According to the principles of statistical mechanics, thermal transport coefficients of fluids are generally expressed as integrals of time correlation functions in linear response theory (Green–Kubo formalism) (McQuarrie 1973). The shear viscosity η is given by the time integral

$$\eta(T) = \frac{V}{3k_B T} \int_0^{\infty} \left\langle \sum_{x<y} \sigma^{xy}(t)\sigma^{xy}(0) \right\rangle dt \qquad (10.1)$$

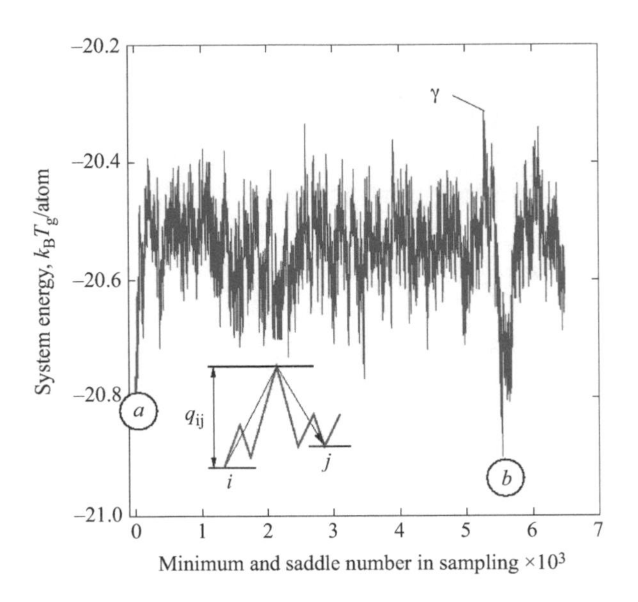

Figure 10.3
TSP trace generated from ABC sampling. Activation barrier q_{ij} is the lowest saddle point between energy minima i and j. T_g is 0.323 in reduced unit. Reprinted from (Kushima et al. 2009a), with the permission of AIP Publishing.

where V denote the volume of the simulation cell maintained at temperature T, and the shear stress tensor is

$$\sigma^{xy} = \frac{1}{V}\left[\sum_j m_j \underline{v}_j^x \otimes \underline{v}_j^y + \frac{1}{2}\sum_{i \neq j} \underline{r}_{ij}^x \otimes \underline{F}_{ij}^y\right] \tag{10.2}$$

where F_{ij} is the force between atoms i and j, r is the position vector and v is the velocity. For simplicity, we will suppress the superscript xy and introduce the time-dependent stress autocorrelation function $S(t) = < \sigma(t)\sigma(0) >$ (recall essay 1). Because in dense gases and normal liquids, $S(t)$ decays on the microscopic timescale of molecular collisions, it can be readily evaluated by MD simulation (Fernandez, Vrabec, and Hasse 2004). In the supercooled regime, the relevant dynamics is believed to be system activation over deep energy minima. Then $S(t)$ no longer decays on the timescale of intermolecular collisions and MD becomes ineffective. To overcome this problem, we invoke the activated-state kinetics involving the TSP trajectories just discussed.

We consider two different formulations, both making use of the TSP trajectory input, to calculate the viscosity in the 1–10^{12} $Pa \cdot s$ range. Despite their appearances, one can be regarded as a simplification of the other (see equation [10.8] in the next section).

A Network Model

In standard linear response theory one can analyze $S(t)$ by treating basin hopping as a random walk on a Markov network of nodes. Imagine that the system is able to sample a number of deep minima within the timescale of the calculation, so the average macroscopic properties like viscosity will take on steady state values. We also assume the activation barriers are high compared to k_BT, and the energy dissipation is efficient so that after each activation the system effectively loses memory of its previous history. Under these conditions the hopping rate from nodes i to j is given by transition state theory (TST)

$$a_{ij} = \nu_o \exp(-Q_{ij}/k_BT) \tag{10.3}$$

where ν_o is a characteristic attempt frequency. One can show (Li et al. 2011) that in this model, the stress correlation function becomes

$$<\sigma(t+\tau)\sigma(t)> = \sum_i P_i\sigma_i g_i(\tau) \tag{10.4}$$

where P_i is probability the system is on node i, σ_i is the shear stress projected on node i, and $g_i(\tau)$ is the average value of the shear stress at time τ given that the system has hopped to node i at time $t = 0$. By taking advantage of the assumed Markov nature of the hopping process, an integral equation can be set up for $g_i(\tau)$ and solved without any approximation

$$g_i(\tau) = \int_0^\tau d\tau' s_i(\tau') \sum_j a_{ij}\, g_j(\tau-\tau') + s_i(\tau)\sigma_i \tag{10.5}$$

where $s_i(t) = \exp(-a_it)$ is the probability that the system will stay at node i during time t, and $a_i = \sum_j a_{ij}$ is the rate at which the system will leave node i. To calculate the viscosity, we then need only the time integral of the correlation function, as indicated in equation (10.1). The result is

$$\eta(T) = \frac{\Omega}{k_BT}\sum_i P_i\sigma_i \frac{(A(\omega=0^+)^{-1}\sigma)_i}{a_i} \tag{10.6}$$

where $A(\omega)$ is the matrix $A_{ij} = \delta_{ij} - \dfrac{a_{ij}}{\omega + a_i}$. Equation (10.6) gives the viscosity in terms of the nodal activation energy Q_{ij}, the energy of the nodes, E_i, for determining $P_i = \dfrac{\exp(-E_ik_BT)}{\sum_j \exp(-E_{j/k_B}T)}$, and the stress at node i,

$$\sigma_i = \frac{1}{V} < Nk_B TI + \sum_{n=1}^{N} x_n \otimes \partial_{x_n} \Phi >_i \tag{10.7}$$

Consider the special case of a two-state model, where $P_1 = P_2 = 1/2$, $\sigma_1 = -\sigma_2 = \sigma_o$, $a_{12} = a_{21} = v_0 \exp(-Q/k_B T)$. Then equation (10.6) reduces to

$$\eta = \sigma_0^2 v_o^{-1} \exp(Q/k_B T) \tag{10.8}$$

The Effective Activation Energy

Equation (10.8) has the form of a widely used expression to correlate experimental data on viscosity (Brush 1962)

$$\eta(T) = \eta_0 \exp[\bar{Q}(T)/k_B T] \tag{10.9}$$

First used by Andrade as a two-parameter formula to describe liquids, the prefactor η_o and the activation barrier $\bar{Q}$ were treated as fitting constants (Andrade 1930). In contrast, we will adopt equation (10.9) as a temperature-dependent relation between the viscosity and an effective activation barrier to be extracted from appropriate TSP trajectory data. The analysis to obtain $\bar{Q}(T)$ is fully discussed in the original publication (Kushima et al. 2009a). Equation (10.9) thus becomes a method to estimate $\eta(T)$ given the effective activation energy. Because $\bar{Q}(T)$ is extracted from the TSP, it can be regarded as the single path approximation (SPA). Henceforth, equations (10.6) and (10.9) become two complementary formulations. They are both predictive in that no experimental input is used in their evaluations. On the other hand, in order to compare the numerical results with MD simulations or experimental data, as will be discussed in the following sections, a normalization is customarily required, simply because TST has been assumed. This is a standard procedure amounting to choosing a value for v_o in equation (10.3), or η_o in equation (10.9).

From our introduction of the network model, equation (10.9) becomes an approximation where viscous relaxation is described by a single effective barrier. It will be seen in the following $\bar{Q}(T)$ has a thermodynamic component arising from the temperature variation of the average inherent structure, as well as a kinetics component associated with the detailed activation barrier analysis.

The connection between the network model and SPA is clearly brought out by equation (10.8), which shows the former reduces to the latter when there is only one activation path. In other words, SPA is a simplification of the network model by ignoring all the correlations (coupling effects) among the different activation paths. One can think of another way to correct the SPA formulation, namely, to take into account entropy effects. This may be implemented by modifying the activation barrier $\bar{Q}(T)$ in equation (10.9) to introduce a degeneracy factor $G(T)$ in the form of an Adam–Gibbs relationship (Heuer 2008)

$$\eta(T) = \eta_0 \exp[\bar{Q}(T)/\{k_B T \log G(T)\}] \tag{10.10}$$

with $G(T)$ being estimated from the density of states $G(E_i)$ distribution of the binary Leonard-Jones (BLJ) model given by Heuer. The effect of this correction will be discussed along with the implementation of equation (10.6).

As mentioned, our viscosity formulations will be tested in two ways. The first is to compare results with Green–Kubo MD, which is feasible only at low-viscosity values. Nonetheless, this is worthwhile because it constitutes a consistent benchmark in which the same interatomic interaction model is used. The second test is to compare results with experimental data in the intermediate- to high-viscosity range ($1-10^{12}\ Pa \cdot s$). In this comparison, we need to keep in mind that our results are obtained using BLJ model potential, which does not necessarily describe any real liquid.

Fragile Viscosity Behavior

At temperatures in the normal liquid range the viscosity values are low enough for conventional MD to be effective in determining $\eta(T)$ from equation (10.1). We have performed standard MD simulations using a periodic cubic simulation cell containing 500–2,048 atoms. After equilibration is attained, typically in 100,000 time steps, the simulation is allowed to proceed for a buffer period of 10,000 iterations without any temperature control. The stress autocorrelation function is then averaged over 2,000 steps in the NVE ensemble and further averaged over 8–12 independent runs.

Figure 10.4 shows the MD results with the viscosity and the temperature expressed in reduced units appropriate to the BLJ potential. In implementing equation (10.6), we take each local minimum given by the TSP trajectory in figure 10.3 to be the energy of a node, and the transition frequency a_{ij} in equation (10.3) is then specified by the activation barrier q_{ij} as labeled in figure 10.3. To put the network model and SPA results in the same form for comparison with MD, we choose to normalize the viscosities at $T = 1$.

Figure 10.4 shows the network model matches well with the MD data. We regard this as perhaps the most critical result of our benchmarking, a confirmation that our formulation of the Green–Kubo formalism to incorporate activated state kinetics indeed captures the temperature variation of the viscosity determined by MD (in the region where the latter is expected to be valid). Notice that the SPA formulation describes a viscosity increase with lowering temperature that is too strong compared to the MD and the network model. This

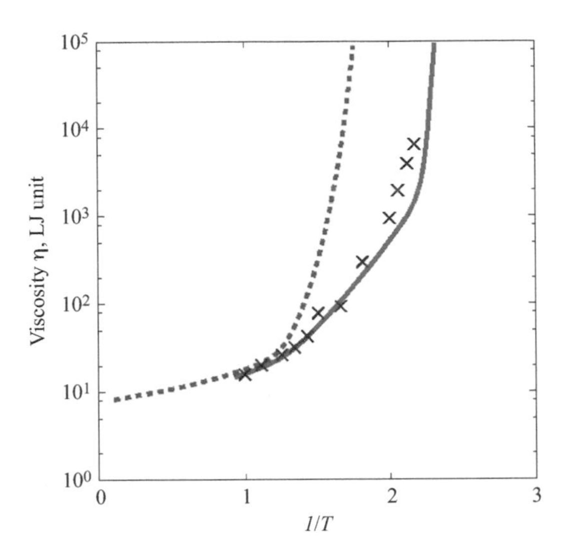

Figure 10.4
Comparison of the shear viscosity calculated using the network model (solid curve) and SPA formulation (dashed curve) with results obtained by MD simulation (crosses). All results are expressed in reduced units appropriate to the BLJ interatomic potential. Reprinted from (Kushima et al. 2009a), with the permission of AIP Publishing.

can be understood as a direct consequence of our activation barrier analysis where $\bar{Q}(T)$ has been obtained as an upper-bound estimate. We can therefore attribute the difference between the network model and SPA results as the effects of activation path coupling that are included in the former but not in the latter. These observations, seen in the low-viscosity region, should also hold in the high-viscosity region where we next test our formulations against experimental data on fragile liquids.

Note that the above derivation starts from Green–Kubo theory and assumes the system is in an equilibrium condition. Calculations presented in this essay are performed under this assumption. However, the network model can be extended to nonequilibrium conditions (Mauro et al. 2007).

Implementation

The fundamental assumption of the present model is that the relaxation kinetics of a supercooled liquid can be described by TSP trajectories in which the system hops from node to node by going over saddle-point atomic configurations. Each trajectory, denoted as $\Gamma(E_m, E_s, \Lambda)$, is an alternating sequence of local energy minima, E_m, and saddle point energies, E_s. Λ represents the collection of indices which specify how the trajectory is generated (sampled), the initial and final states of the trajectory, the system size and preparation (temperature at which the liquid is quenched), and state variables external stress or strain, and so on. We are not concerned with the details of Λ except to note that trajectories sampled under different conditions can provide specific details about the system energy landscape, so that the calculation of $\eta(T)$ over a significant range of values should involve a distribution of contributions from one or more TSP trajectories as input.

To be more specific, we recall how a system, with a prescribed interatomic interaction potential, is prepared for TSP trajectory sampling. We start with a periodic simulation cell with N particles and an appropriate thermostat for MD simulation. After the system is equilibrated in the liquid state, it is quenched to a temperature T_f below the melting point. MD simulation is continued at T_f during which a series of steepest descent relaxations is performed to obtain a distribution of energy minima (the inherent structure) and corresponding atomic configurations (Sastry, Debenedetti, and Stillinger 1998). From this distribution, the initial states for TSP trajectory sampling are selected. Each trajectory, generated by an activation-relaxation algorithm, therefore can be characterized by the temperature T_f and the state, local minimum energy, and associated atomic configuration, where the trajectory is initialized.

Figure 10.5 shows the connection between the description of potential well depths based on the notion of inherent structure of a supercooled liquid and the description of TSP in the form of trajectories generated (sampled) by the metadynamics activation-relaxation algorithm. The results are obtained for a model binary liquid where particles interact through a particular Lennard-Jones interatomic potential (Li et al. 2011). In the left panel, figure 10.5 shows the temperature variation of the average inherent structure $\bar{E}_{\text{IS}}(T)$, which we may interpret as the average well depth of the local minima that the system would encounter at temperature T. At high temperatures, $1/T < 1.0$, one sees shallow wells, which is physically self-evident. As temperature decreases, $1.0 < 1/T < 3$, the wells become deeper until $1/T > 3$ when maximum depth is encountered. Corresponding to the behavior of the average inherent structure is the distribution of well depths at a given temperature, seen in the central panel. This distribution is seen to be very broad at high temperatures and very narrow at low temperatures; the sharpening of the distribution occurs rather suddenly in the range $0.4 < T_f < 0.3$, not far from the inflection point in $\bar{E}_{\text{IS}}(T)$. In contrast to the temperature dependence of the distribution and average depth of the local minima, the right panel in figure 10.5 shows representative TSP trajectories sampled at several initial states in the energy landscape. Recall from the above discussion that a trajectory is labeled by $\Gamma(E_m, E_s, \Lambda)$, where Λ includes the specification of the initial state of the trajectory E_o. Figure 10.5 shows the trajectories obtained at the same value of

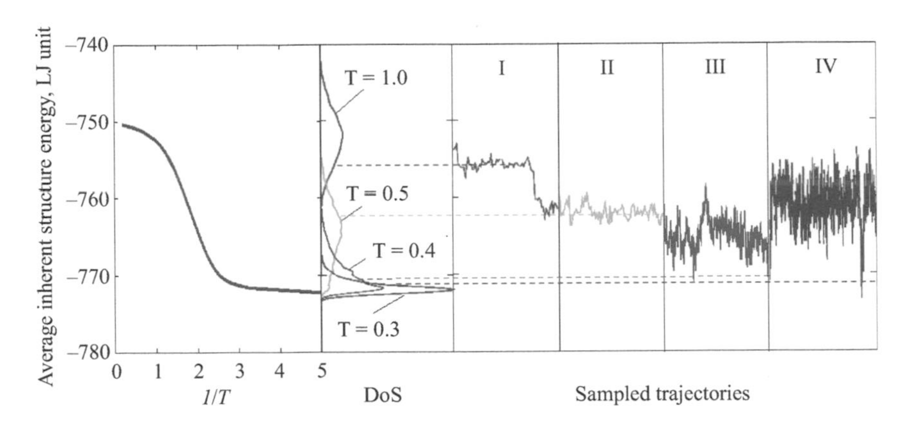

Figure 10.5
Average inherent structure (IS) energy of BLJ liquid as a function of temperature (left panel), quench probability distribution at temperatures 1.0, 0.5, 0.4, and 0.3 (middle panel), and four TSP trajectories initialized at different energy minima (right panel). Figure from (Li et al. 2011).

$T_f = 0.5$ but four different values of E_o. It is clear that the results are sensitive to E_i, implying the sampling is apparently confined to a rather small portion of the energy landscape around E_o. From this observation, it follows that to achieve a reasonably averaged calculation of $\eta(T)$ over a wide temperature range, several trajectories will be needed to cover a sufficient portion of the energy landscape that would be sampled by the system.

Considering the four trajectories in figure 10.5 in detail, we see in (I) that E_o lies near the top of the distribution range, local minima E_m are shallow, and the saddle point energies are low. Similarly in (II), E_o lies near the peak of the distribution, E_m is somewhat deeper, and saddle points relatively larger. In (III), E_o is lower than in (II), E_m is deeper, and saddle point (activation) energies are larger. In (IV), E_o is still lower, E_m can reach still deeper values, and saddle points energies significantly larger. Based on these qualitative features, we expect the temperature range where a given trajectory may be relevant for the calculation of $\eta(T)$ to be the following: Trajectory (I) $1/T < 1.5$, (II) $1.5 < 1/T < 2$, (III) $2 < 1/T < 2.5$, (IV) $1/T > 2.5$. In other words, we expect (I) and (II) to give reasonable results for $\eta(T)$ at high temperatures, (IV) should be useful at low temperatures, while (III) can describe the transition between high and low temperatures. Before showing the results of $\eta(T)$ obtained using these trajectories, we note the number of minima sampled in the four trajectories are 70, 80, 480, and 3,000, respectively. This reflects our belief that larger number of local-minima sampling is needed to escape from a deep well and to fall into another deep well (Li et al. 2011).

Figure 10.6 shows the viscosities in a certain temperature range calculated using the four trajectories in figure 10.5, with different symbols for each trajectory. The choice of

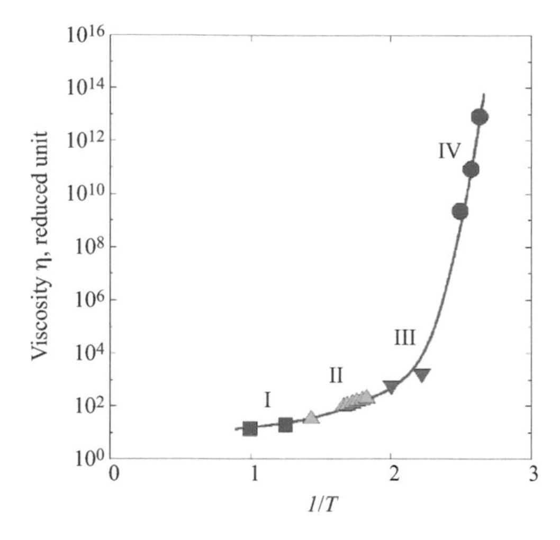

Figure 10.6
Viscosity computed using the network model expression with the four TSP trajectories shown in figure 10.5 as input. Results for each trajectory are denoted by a different symbol, squares for trajectory I, triangle for II, inverted triangles for III, and circles for trajectory IV. The solid curve is a spline fit to all the calculated viscosities. Figure from (Li et al. 2011).

temperature range in each case follows the reasoning just discussed. The curve in figure 10.6 is a fit to all the data points using a cubic spline. It can be seen that the results given by trajectories (I) and (II) connect smoothly with each other, and the trend extends to the first point from trajectory (III). Except for the second point from (III), all the calculated viscosities lie well on the fitted curve. We believe the underestimate by this second point could be an indication of insufficient sampling of the activation kinetics, 480 minima sampled in (III) compared to 3,000 in (IV). Given the results presented, we think that the fitted curve represents an optimum estimate of $\eta(T)$ over the entire temperature range of interest, calculated using an interatomic potential model and without using any input from experiments. This is the first goal of the present work. Now we address the issue of further testing the results of the network model.

Validation against Molecular Dynamics Simulations

We have two ways of assessing the validity of the viscosity results shown in figure 10.4. One is by comparison with MD simulations carried out using the same interatomic interaction potential. This would lead to a self-consistent demonstration that the network model formulation gives the same results as the traditional Green–Kubo method. A second way is to directly compare the calculated results for the BLJ potential with

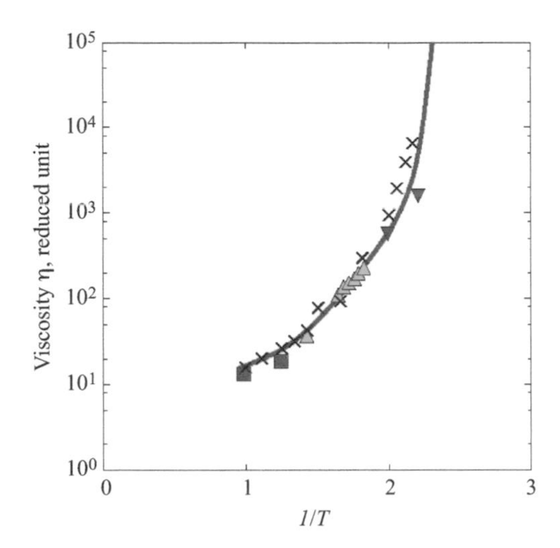

Figure 10.7
Comparison of network model result (solid line) against Green–Kubo MD (crosses). Figure from (Li et al. 2011).

experimental data on liquids that display fragile temperature scaling. While this necessarily involves an ambiguity in that the BLJ potential is not designed to describe accurately any real liquid, it is nevertheless perhaps the most direct way to ascertain whether the characteristic fragile behavior observed experimentally can be explained a priori by an atomistic calculation.

At temperatures near the melting point, the viscosity values of most liquids are low enough for using MD and Green–Kubo formalism to obtain reliable results for $\eta(T)$ (Grigera et al. 2002). We have performed standard MD simulations using periodic cubic simulation cells containing 500–2,048 atoms. After equilibration is attained, typically in 100,000 time steps, the simulation is allowed to proceed for a buffer period of 10,000 iterations without any temperature control. The stress autocorrelation function is then averaged over 2,000 sets in the NVE ensemble and 8–12 independent runs.

Figure 10.6 shows the comparison of network model (same symbols as in figure 10.7 below) and MD (shown as crosses) results in the region of relatively low viscosity. The same BLJ potential model is used in the two independent calculations. The network model calculation requires the specification of the prefactor v_0 in equation (10.3). We choose a value of v_0 such that the network model and the MD results agree in the high-temperature limit (taken to be $T = 1$). It is seen in figure 10.7 that the two methods give results agreeing well with each other over three decades of variation. We regard this to

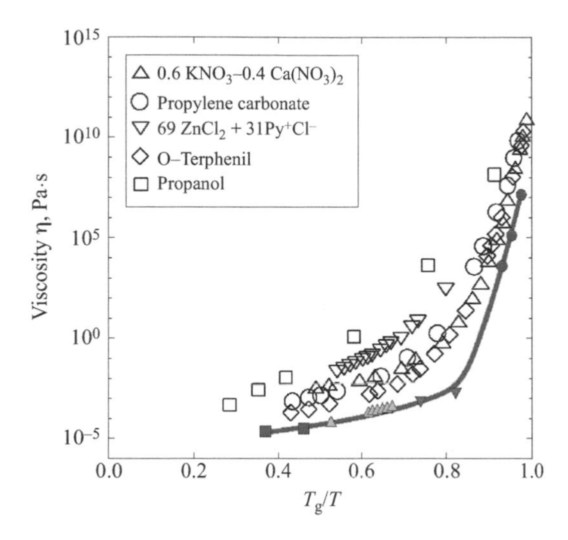

Figure 10.8
Experimental validation of the network model. Solid line indicates the viscosity of BLJ liquid calculated by the network model. Symbols are experimental data of fragile glass formers. Figure from (Li et al. 2011).

be the direct evidence that our extension of the Green–Kubo formalism to activated state kinetics is correctly developed, and that the use of appropriate TSP trajectories as input to the network model to describe $\eta(T)$ is also correct.

Validation against Experiments

A particular challenge in testing $\eta(T)$ calculation is the region of high viscosities. Figure 10.8 shows the results of the network model in the macroscale range of ~10^{13} in reduced units (or 10^8 Pa·s). Since no MD results are available in this region, we resort to assessing the network model results by direct comparison with experiments, keeping in mind that the BLJ potential is not designed for any single actual liquid. The prefactor ν_0 is specified by the condition that at high temperature, the viscosity extrapolates to a value of 10^{-5} Pa·s, a behavior seen in many experimental data. The network model results based on this normalization are shown in figure 10.9, along with a collection of data on five liquids that are generally considered to display representative fragile behavior in their temperature variation. All viscosities are given in absolute unit and temperature is scaled in T_g, defined as $\eta(T_g) = 10^{12}$ Pa·s. For the BLJ potential model, we have previously determined T_g to be 0.37 in reduced unit. While there are no predictions of T_g with which we can compare, in the mode-coupling theory of ideal glass transition, the critical temperature T_c has been estimated for the BLJ potential model to be $T_c = 0.435$ (Debenedetti and Stillinger 2001). This gives a ratio $T_c/T_g \sim 1.3$, which is consistent with general expectations (Coslovich and Pastore 2007).

Our primary concern here is to assess how well the network model can account for the characteristic non-Arrhenius behavior seen in the data. In figure 10.8, we would regard the agreement between the network model and the experimental trend to be satisfactory, given the ambiguity of any direct comparison between measurement on real liquids with simulation results based on the BLJ model. Specifically, the calculated temperature variation from the onset of fragile behavior (the knee of the curve) to the region around T_g is seen to follow the outer envelope of the five-liquid data set in a semiquantitative fashion.

Another form of experimental validation is to express viscosity and relaxation time data in terms of an effective temperature-activation barrier (Stillinger and Weber 1982). The idea of focusing on an activation barrier is similar to the heuristic treatment we had previously proposed in which a coarse-grained activation barrier $\bar{Q}(T)$ was extracted from the TSP trajectory shown in figure 10.3. The resulting barrier shows a simple variation with T beginning at a constant, $\bar{Q}_\infty$, at high temperatures. With decreasing temperature, $\bar{Q}(T)$ remains constant until T reaches a characteristic value T^*, when $\bar{Q}(T)$ begins to increase sharply. If we plot the quantity $(\bar{Q}(T) - \bar{Q}_\infty)$ versus $(T^*-T)/T^*$, we would see a behavior very similar to the plot of $k_B T \ell n[\eta(T)/\eta_\infty]$ versus T_g/T. In other words, one can collapse the experimental data in figure 10.8 into a master curve by appropriate scaling. Figure 10.9 shows such a reduction applied to a group of fifteen liquids (Kivelson et al. 1996). For comparison, we show that the network model results reduced similarly. In this way of comparing theory and experiments, fragile behavior becomes the onset of temperature-sensitive activation around the characteristic temperature T^*. Here we are able to relate T^* to T_c in the mode coupling theory. The extent of the agreement over the range of comparison indicated in figures 10.6 and 10.7 forms a basis for asserting a description of ultra-viscous liquids where previously no molecular theory existed is now demonstrated.

The agreement in the high-barrier (low-temperature, below T_c) region is particularly noteworthy. It suggests a universal behavior that is brought out more fully by the network model than the heuristic model (dashed curve). As indicated in our discussion, the latter is a single effective activation path approximation. Therefore, the coupling effects ignored must play a significant role. A discernible discrepancy between the network model and the experimental trend remains in the lower barrier region (temperature above T_c). This is an overestimate of the effective activation, which may be attributed to the absence of

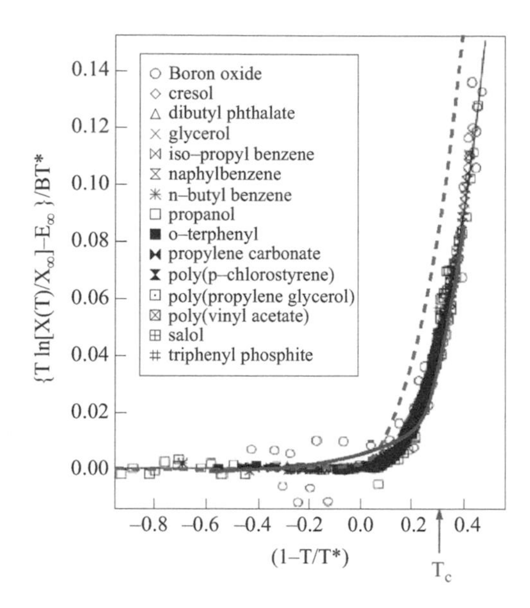

Figure 10.9
Experimental test of the activation barrier obtained from the temperature dependence of the BLJ liquid calculated by the heuristic model and the network model of figure 10.5 against scaled activation barriers derived from viscosity measurements (symbols). Except for scaling the axes, no adjustable parameters are involved in this comparison. Figure from (Li et al. 2011).

entropic effects or an effect of inaccurate potential model. Because this is not an underestimate, one can rule out insufficient TSP sampling in the trajectories actually used as a possible cause.

Discussion

In this essay, we demonstrate an alternative method of molecular dynamics (MD) simulation, one that can reach much longer time scales. This is accomplished by showing that the well-known fragile behavior of the viscosity of supercooled liquids can be obtained through modeling and simulation at the temporal mesoscale level. A companion part of the story concerns understanding the nature of the liquid-to-glass transition. This discussion is taken up in the following essay. Together, the two essays provide an introduction to the metadynamics approach to probing molecular mechanisms in amorphous matter, a topic central to the frontiers of computational materials science.

It is an advantage of the molecular simulation method that one can use appropriate interatomic interaction potentials to study different physical systems (recall essay 2).

For example, we can apply the present meta-dynamics method based on TST to study a typical strong glass former. This will be demonstrated in the following essay, the companion to essay 10. Another advantage is the availability of atomic configurations that can provide insights into mechanisms of activation and relaxation. Consider how the atomic configurations at different points along a TSP could be correlated and displayed, an example being figure 9 in Kushima (2009a), which shows the atomic configurations at the three points indicated as (a), (b), and (c), corresponding to activation events at low-energy, intermediate-energy, and high-energy, respectively. The insight gained is that a low-activation event involves only a few atoms, while at higher activation, chain-like collective displacements can take place, and at even higher activation, particle displacements become chaotic again. In other words, chain-like cooperative displacements are best probed among intermediate activation-energy events. By extension, it implies there is an optimum timescale for organized cooperative rearrangements. While our remarks are by no means definitive, they are suggestive and consistent with recent investigations of the mechanisms of hopping between PES metabasins (Appignanesi et al. 2006; Doliwa and Heuer 2003). Insights of this kind, when combined with topographical features of the TSP trajectory discussed in figure 10.5, could lead to broader understanding of the kinetics of slow system evolution.

In summary, we have examined a TSP trajectory sampling algorithm used in conjunction with two complementary methods of calculating the viscosity of highly viscous liquids. Of paramount concern throughout is the validity of the basin-filling algorithm, the activation barrier analysis, and the formulations to calculate viscosity in the high-viscosity range. These issues have been addressed at three stages. First the algorithm was benchmarked by applying it to the problem of ad-atom diffusion on a metal surface (not discussed here) (Kushima et al. 2009a). Then, the two different formulations for computing the viscosity, the network model and SPA are compared with MD simulations in the low-viscosity region (shown in figure 10.7), and also with experimental data in the high-viscosity region, (shown in figure 10.8). In both cases, the network model is found to be quite accurate, while the SPA is able to account for the characteristic fragile scaling behavior in a qualitative manner. Additionally, the coarse-grained temperature-dependent activation barrier $\bar{Q}(T)$ was tested against experimental results in figure 10.9.

In the essay to follow, the heuristic approach is applied to a potential model for SiO_2, silica, a system with measured viscosity that follows closely an Arrhenius behavior (Kushima et al. 2009b). The combination of studies of fragile and strong behavior allows further insights into the nature of the liquid-to-glass transition. Our findings suggest fragility, with its fundamental connection to current understanding of the dynamics of glass transition (Debenedetti and Stillinger 2001; Dyre 2006; Trachenko and Brazhkin 2008), is a universal manifestation of all systems undergoing slow deformational changes. We believe that there is no qualitative distinction between strong versus fragile liquids—even the so-called strong liquids should show a fragile-to-strong transition, albeit at high temperature and over a smaller range. Such a point of view has been raised from the perspective of spatial scales in glassy systems (Trachenko and Brazhkin 2008).

References

Andrade, E. N. da C. 1930. "The viscosity of liquids." *Nature* 125: 309.

Angell, C. A. 1995. "Formation of glasses from liquids and biopolymers." *Science* 267: 1924.

Appignanesi, G. A., J. A. Rodriguez Fris, R. A. Motani, and W. Kob. 2006. "Democratic particle motion for metabasin transitions in simple glass formers." *Physical Review Letters* 96 : 057801.

Brush, S. G. 1962. "Theories of liquid viscosity." *Chemistry Review* 63: 513.

Coslovich, D., and G. Pastore. 2007. "Understanding fragility in supercooled Lennard-Jones mixtures. II. Potential energy surface." *Journal of Chemistry and Physics* 127: 124505.

Debenedetti, P., and F. Stillinger. 2001. "Supercooled liquids and the glass transition." *Nature* 410: 259.

Doliwa, B., and A. Heuer. 2003. "Energy barriers and activated dynamics in a supercooled Lennard-Jones liquid." *Physical Review E* 67: 031506.

Dyre, J. C. 2006. "Colloquium: the glass transition and elastic models of glass-forming liquids." *Reviews of Modern Physics* 78: 953.

Fernandez, G. A., J. Vrabec, and H. Hasse. 2004. "A molecular simulation study of shear and bulk viscosity and thermal conductivity of simple real fluids." *Fluid Phase Equilibria* 221: 157.

Grigera, T. S., A. Cavagna, I. Giardina, and G. Parisi. 2002. "A molecular simulation study of shear and bulk viscosity and thermal conductivity of simple real fluids." *Physical Review Letters* 88: 055502.

Henkelman, G., and H. Jonsson. 1999. "Improved tangent estimate in the nudged elastic band method for finding minimum energy paths and saddle points." *Journal of Chemistry and Physics* 111: 7010.

Heuer, A. J. 2008. "Exploring the potential energy landscape of glass-forming systems: from inherent structures via metabasins to macroscopic transport." *Physics: Condensed Matter* 20: 373101.

Kivelson, D., G. Tarjus, X. Zhao, and S. A. Kivelson. 1996. "Fitting of viscosity: distinguishing the temperature dependences predicted by various models of supercooled liquids." *Physical Review E* 53: 751.

Kob, W., and H. C. Andersen. 1995. "Testing mode-coupling theory for a supercooled binary Lennard-Jones mixture I: the van Hove correlation function." *Physical Review E* 51: 4626.

Kushima, A., X. Lin, J. Li, J. Eapen, J. C. Mauro, X. Qian, P. Diep, and S. Yip. 2009a. "Computing the viscosity of supercooled liquids." *Journal of Chemistry and Physics* 130: 224504.

Kushima, A., X. Lin, J. Li, X. Qian, J. Eapen, J. C. Mauro, P. Diep, and S. Yip. 2009b. "Computing the viscosity of supercooled liquids. II. Silica and strong-fragile crossover behavior." *Journal of Chemistry and Physics* 131 (16): 164505.

Laio, A., and M. Parrinello. 2002. "Escaping free-energy minima." *Proceedings of the National Academy of Science* 99: 12562 .

Li, J., A. Kushima, J. Eapen, X. Lin, X. Qian, J. C. Mauro, P. Diep, and S. Yip. 2011. "Computing the viscosity of supercooled liquids: Markov network model." *PLos ONE* 6 (3): e17909.

Löwen, H. 1994. "Melting, freezing and colloidal suspensions." *Physics Report* 237: 249.

Martoňák, R., A. Laio, and M. Parrinello. 2003. "From four-to six-coordinated silica: transformation pathways from metadynamics." *Physical Review Letters* 90: 075503.

Massen, C. P., and J. P. K. Doye. 2007. "Power-law distributions for the areas of the basins of attraction on a potential energy landscape." *Physical Review E* 75: 037101.

Mauro, J. C., R. J. Loucks, J. Balakrishnan, and S. Raghavan. 2007. "Monte Carlo method for computing density of states and quench probability of potential energy and enthalpy landscapes." *Journal of Chemistry and Physics* 126: 194103.

McQuarrie, D. A. 1973. *Statistical Mechanics*. New York: Harper and Row.

Sastry, S., P. Debenedetti, and F. Stillinger. 1998. "Signatures of distinct dynamical regimes in the energy landscape of a glass-forming liquid." *Nature* 393: 554.

Stillinger, F. 1995. "A topographic view of supercooled liquids and glass formation." *Science* 267: 1935.

Stillinger, F., and T. A. Weber. 1982. "Hidden structure in liquids." *Physical Review A* 25: 978.

Trachenko, K., and V. V. Brazhkin. 2008. "Hidden structure in liquids." *Journal of Physics: Condensed Matter* 20: 075103.

Wang, F., and D. P. Landau. 2001. "Efficient, multiple-range random walk algorithm to calculate the density of states." *Physical Review Letters* 86: 2050.

Further Reading

Kob, W. 1999. "Computer simulation of supercooled liquids and glasses." *Journal of Physics: Condensed Matter* 11: R85. Comprehensive review of advantages and drawbacks of MD simulation of the dynamics of supercooled liquids with discussions of both fragile and strong glass formers as well as the validity of the mode-coupling theory in the context of an ideal glass transition (in the absence of barrier activation).

Scalliet, C., B. Giuslin, and L. Berthier. 2022. "Thirty Milliseconds in the Life of a Supercooled Liquid." *Physical Review X* 12: 041028. Relaxation dynamics of model supercooled liquids simulated by molecular dynamics over a time range of 10 orders of magnitude.

11 Glass Transition

This is the companion to essay 10 where we discuss a metadynamics simulation approach in the case study of the viscous behavior of supercooled (glass-forming) liquids. We have seen from a considerable body of data that there are two types of temperature variations of the shear viscosity of these liquids: one is exponential in inverse temperature, or Arrhenius, the other is more extreme, super Arrhenius. The former is what would be considered normal behavior for transport properties of fluids, whereas the latter, the so-called *fragile* behavior, is the more exceptional and therefore the subject of interest regarding the nature of the liquid-to-glass transition. The case of Arrhenius behavior has been dubbed *strong* in contrast.

From the standpoint of physical understanding, one would like a conceptual framework in which both fragile and strong behavior can be explained in a self-consistent framework. Thus, it is not enough to explain the fragile behavior of super-cooled liquids alone without due considerations given to the strong liquids. In this essay, we apply the same transition pathway sampling approach to a particular example of the latter, to find similar accountability of the experimental data. On this basis, we then claim that the characteristic kinetics of glassy relaxations can be explained.

Introduction

We discuss an approach to characterize the viscous behavior of super-cooled liquids using molecular simulations capable of reaching timescales beyond those of traditional MD. The problem has universal interest as the signature of the liquid-to-glass transition; it is a longstanding challenge in the nonequilibrium statistical mechanics community, among others. Experimentally, the shear viscosity of glass-forming liquids cooled below freezing shows anomalous variations with the quench temperature. Two types of behavior are observed: Arrhenius and highly super-Arrhenius. Conceptually, the molecular description of the structural and dynamical processes underpinning the viscosity behavior has become a topic of interest because of general implications for the kinetics of slow relaxation in the glassy state.

We regard the super-cooled liquid as an assembly of interacting particles undergoing thermal fluctuations such that the system evolves in space and time by crossing a series of potential-energy barriers. To find these barriers and their corresponding atomic configurations, we apply an energy-landscape sampling algorithm that maps out the evolution trajectories in the form of alternating sequences of local energy minima and saddle points. The energy sequences and the corresponding atomic coordinates constitute a body of atomic-scale data, which we then process using two distinct methods. One is based on an effective activation barrier extracted from the TSP data, and the other is based on the linear response theory of statistical mechanics. The former is heuristic by being more physically transparent, while the latter is theoretically more rigorous. Through these two complementary calculations, an understanding of the temperature variations of shear viscosities of supercooled liquids, as well as the nature of fragile and strong behavior of the glass transition, emerges.

Our calculation provides a molecular-level account of the viscosities of supercooled liquids in a unified and consistent manner without invoking ad hoc assumptions. Relative to the nature of the glass transition, the usefulness of the potential-energy landscape perspective is demonstrated, along with the concept of crossover between strong and fragile behavior. In terms of advancing atomistic simulation capability, we believe that the timescale limitations of traditional MD may be significantly mitigated through the use of metadynamics algorithms for sampling TSPs.

In this essay, we rephrase our objective in the form of a challenge. Given that the ABC simulation method has been introduced and applied to study the fragile behavior of supercooled liquids (preceding essay), we now focus on the broader issue of explaining *both* fragile and strong behavior *within a consistent theoretical framework.*

The GAS Challenge

In 1969, M. Goldstein proposed that studies of the metastability and transport of supercooled glass forming liquids can benefit from a potential energy landscape perspective in which viscous flow is dominated by activation over barriers on this landscape (Goldstein 1969). He also emphasized the need to focus on molecular mechanisms governing a broader class of kinetic and rate-dependent phenomena such as creep deformation. About two decades later, C. A. Angell summarized the experimental data on shear viscosity of supercooled glass-forming liquids in a way that calls attention to two distinct types of temperature variation, which he labeled as *fragile* and *strong,* and challenged the community to explain them theoretically while embracing the PEL concept (Angell 1988). In the meantime, F. H. Stillinger, who has been concerned with statistical mechanics issues of the glass transition, presented a topographic view that extends the concept put forth by Goldstein (Stillinger 1988; Stillinger 1995; Debenedetti and Stillinger 2001). Taken together, these contributions presented a challenge to the community calling for the understanding of the viscosity of supercooled liquids with its implications for the glass transition and beyond. In this essay, we will regard the ideas of Goldstein, Angell, and Stillinger as a composite challenge, denoted as GAS in what follows (Yip 2016).

With GAS as the context, we proceed to examine a molecular calculation of the viscosity in the spirit of a response to GAS, almost four decades after the Goldstein suggestion. The original work, a series of three calculations (Kushima et al. 2009a; Kushima et al. 2009b; Li et al. 2011), did not particularly emphasize the GAS challenge, nor did a previous commentary (Kushima, Lin, and Yip 2009) and a later perspective (Kushima et al. 2011). As noted in the Introduction, our interest here is to bring out aspects of the molecular calculation directly in the GAS context. In the interest of brevity, technical details will be deferred mostly to the previous reports.

The glass transition has a reputation in the physics, chemistry, and materials science communities for being a ubiquitous phenomenon for which there exists no satisfactory physical explanation based on the principles of statistical mechanics. A glass is clearly not an equilibrium state of matter given its properties are known to change with time, albeit slowly. It is considered a system in metastable state, trapped between two equilibrium states, the crystal and the liquid. Moreover, many physical properties of glasses depend sensitively on the condition under which they are formed, by quenching (supercooling or vitrification) a liquid to below its freezing temperature, or by structurally perturbing a crystal through reaction, deformation or irradiation (solid-state amorphization). The fact that glass properties are history-dependent clearly indicates glasses are different states of matter from crystals and liquids.

The shear viscosity $\eta(T)$ is perhaps the most fundamental property of glass-forming liquids. Its variation with the temperature T to which the liquid is quenched shows a behavior that is the essence of glass formation (vitrification). The viscosity of a liquid is a measure of shear stress relaxation. Physically, it can be regarded as the product of the shear modulus and a structural relaxation time. When the viscosity increases strongly with lowering the quench temperature, the behavior is attributed mostly to a rapidly increasing relaxation time, the characteristic of slow dynamics. Three types of questions arise in connection with this ubiquitous phenomenon. First is conceptual—slow dynamics has implications for the fundamental understanding of the underlying PEL (Goldstein 1969). Second is experimental—how can one explain the observed fragile and strong temperature behavior seen in the experimental data (Angell 1988). Third is both theoretical and computational—how can one sample the PEL of supercooled liquids to directly assess the temperature dependence of their shear viscosity (Stillinger 1988; Stillinger 1995; Debenedetti and Stillinger 2001). Taken together over the period in which these questions have been discussed in the community, we regard them as a single composite challenge that we will refer to as GAS in this essay.

Figure 11.1 is a widely known plot of a body of experimental data which has become the signature figure in many discussions of the glass transition (Angell 1988). Although rather busy in its original appearance, this figure is a useful introduction to our present discussion of molecular simulations. For our purpose, it is sufficient to consider a simplified plot, shown in figure 11.2 (Debenedetti and Stillinger 2001). Both figures show the temperature-dependent $\eta(T)$ in the same format. The experimental data on shear viscosity of several glass formers are expressed in the absolute unit of poise ($1\ p = 10$ Pa·s) on log scale, while temperature T is shown on a linear scale of T_g/T, with T_g defined to be that value of T at which the viscosity of the supercooled liquid has a value of $10^{13}\ p$. It would be appropriate to regard the constant T_g to be the *glass transition* temperature of the particular liquid. However, one should keep in mind that this definition is arbitrary in the sense that other definitions also have been adopted in the literature. However the data are displayed, one has two types of variations with T_g/T in figure 11.1. The data on silica SiO_2, indicate an Arrhenius behavior, $\eta \sim \exp(Q(T)/T)$, with $Q(T)$ being a constant and having the physical meaning of an effective activation barrier. The other is a concave upward behavior, O-terphenyl for example. If the latter is also described by an exponential, then for these liquids, $Q(T)$ would increase strongly with decreasing T.

Figure 11.1 is a master plot summarizing the essential features of the shear viscosity in its variation with the quench temperature for a number of glass forming liquids. By calling attention to two distinct types of behavior, the Arrhenius variation labeled as *strong*, and the other highly super-Arrhenius, labeled as *fragile*, the challenge of understanding the glass transition is separated into two parts (Angell 1988). A fundamental explanation of $\eta(T)$ then should be capable of addressing both variations. To emphasize this observation figure 11.2 shows a simplified version of figure 11.1, where only a few representative data points are displayed (Debenedetti and Stillinger 2001). Figure 11.2 poses a clear challenge to our understanding of transport phenomena in condensed matter. In the fundamental theory of liquid-state transport, one may distinguish two complementary branches. One is focused on phase-space distribution functions as described by the Boltzmann equation with Enskog modification and correlated binary-collision corrections (Mazenko and Yip 1977). The other is a linear response theory approach based on time correlation functions (Martin 1968). In either case, the calculation of $\eta(T)$ for supercooled liquids is a formidable

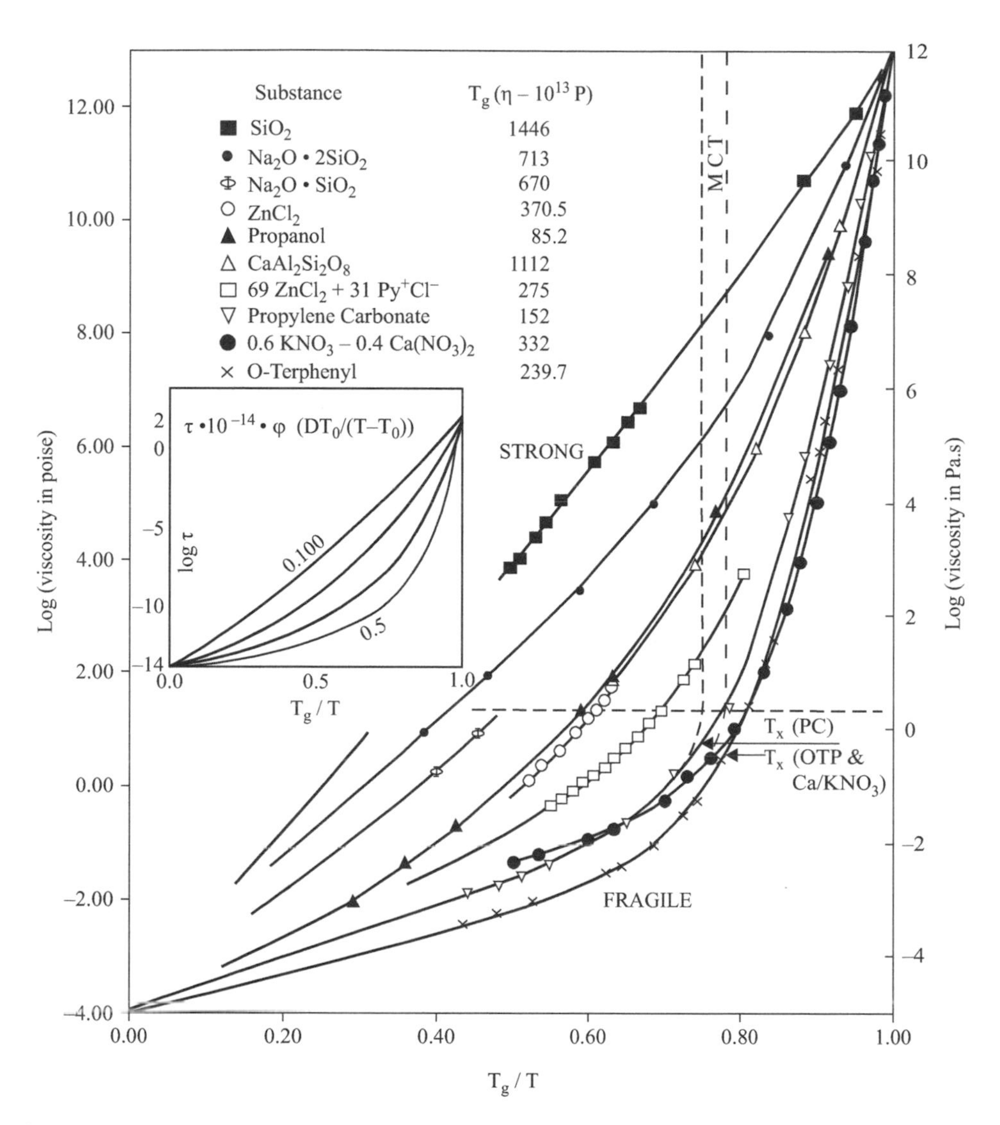

Figure 11.1
Variation of viscosity η of supercooled glass-forming liquids with inverse of quench temperature T reduced by the glass-transition temperature T_g of each liquid. This figure is widely known as the Angell plot. Figure from (Angell 1988).

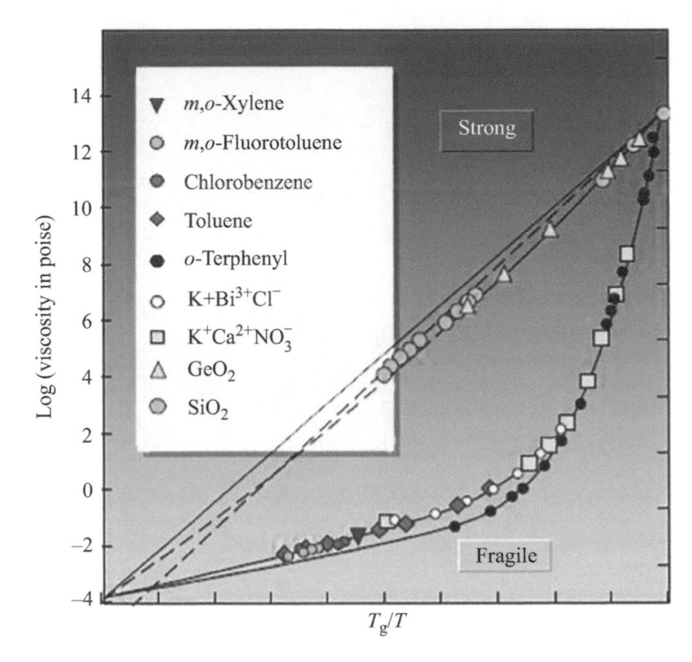

Figure 11.2

Simplified version of figure 11.1 with emphasis on the contrast between *fragile* liquids (the first seven on the list) and *strong* liquids (SiO_2 and GeO_2). This is the essence of the GAS challenge. The range of t_g/T is (0,1.0). Reprinted by permission from Springer Nature: (Debenedetti and Stillinger 2001).

task (Götze 2007; Das 2011). From the standpoint of atomistic simulation, the Green–Kubo formulation is well-suited to MD simulations in which the transport coefficient is given as the time integral of an appropriate time correlation function. However, there are timescale limitations that make it very challenging to calculate $\eta(T)$ over an extended range of quench temperature (Kushima et al. 2009a; Kushima et al. 2009b; Kushima et al. 2009).

When a liquid is quenched to a lower temperature, figure 11.2 shows that $\eta(T)$ can increase very sharply for the fragile liquids after T_g/T has reached a value around 0.8. Considering the shear viscosity as the product of the shear modulus and a relaxation time, one can conclude that the variation is likely to be due to a relaxation time with the modulus being relatively less temperature sensitive. This simple argument states the single most significant dynamical characteristic of supercooled liquids. The same argument also explains why it would be difficult to account for the temperature variations shown in figure 11.2 by molecular simulation. As we have already discussed in the previous essay, in linear response theory, the viscosity is proportional to a time integral of the stress autocorrelation function $S(t)$, or effectively the relaxation time $\tau(T)$. From MD simulation, one can obtain $S(t)$ directly and then $\tau(T)$ by numerical integration. This approach is tenable at high temperature until $\eta(T)$ reaches its rapidly increasing regime (see figure 11.2). When this occurs, direct integration of $S(t)$ will not converge, which makes high-viscosity out of reach for MD.

Transition State Pathways of Supercooled Liquids

A TSP is the trajectory a system follows in moving across its PEL (Wales 2003). The landscape is a $3N$ dimensional surface, N being the number of particles in the system. Suppose that the system moves in discrete steps hopping over a series of potential barriers; its position on the energy surface is determined by the coordinates of all the particles. On an energy surface, these positions are the potential energies of the system. Conceptually, we imagine a cut across the energy surface and refer to it as a certain trajectory on the energy landscape. A TSP is therefore a sequence of energies. See figure 10.3. A simple relation between PEL and TSP is the analogy between a rugged mountainous terrain and the tracks of a solo climber traversing this terrain. There is already a great deal of atomic-level details encoded in a TSP which, however, are minuscule on the multidimensional scale of PEL. Since one can never know the entire PEL, there will always be a concern that estimates of $\eta(T)$ based on limited information gathered by sampling TSP are inadequate. This is where experimental data are useful in assessing the simulation results.

For a given problem, the topographic features of the associated PEL can be deduced by following the system energy whenever it moves across the landscape—a process called sampling. If the system initially is not at a local minimum, it will move to a lower energy state when allowed to relax, such as in energy minimization. Here moving means changes occurring in the atomic coordinates. If the system is initially at a local minimum, it can be activated externally to move away from the minimum and up the potential well. Through a series of external activation and internal relaxation, a portion of the PEL can be determined. Among the different sampling methods that have been proposed, we focus on the ABC method, originally developed for sampling TSPs in supercooled liquids (Kushima et al. 2009a; Kushima et al. 2011). ABC has been used subsequently for a number of other studies (Fan et al. 2013). Additionally, the basic algorithm has been extended (Fan, Yip, and Yildz 2014) and modified to be more efficient (Cao et al. 2012). For the use of ABC in studying fragility in the viscosity of viscous liquids, see the preceding essay.

Temperature-Dependent Effective Activation Barrier of Silica

In motivating the analysis of strong viscosity behavior, we recall that the basic content of TSP is the connectivity between any arbitrary pair of local energy minima. By connectivity, we mean the series of saddles and minima lying between these two states. A coarse-graining procedure that can condense all the statistical variations into a single quantity, an effective temperature-dependent activation barrier, therefore could be what Goldstein had in mind (Goldstein 1969). For such analysis, we invoke an expression from TST (Eyring 1932; Wigner 1932; Laidler 1983)

$$\eta(T) = \eta_0 \exp[\bar{Q}(T)/k_B T] \quad (11.1)$$

This relation was first used by Andrade to fit the viscosities of liquids, with the prefactor and the activation barrier treated as adjustable constants (Andrade 1930). Although clearly an approximation in contrast to equation (10.1), equation (11.1) shifts the understanding of $\eta(T)$ variation over many decades to explaining the variation of an effective activation barrier, $\bar{Q}(T)$ over a much more reduced range of temperature (see essay 10 for further discussions).

A little reflection shows that equation (11.1) quite naturally leads to a strategy for exploiting the state-to-state connectivity encoded in the TSP. In analyzing the data on activation pathways in TSP it will be useful to consider a corresponding network model of interconnected sites, where each site denotes a local energy minimum and the site-site transitions are governed by the activation energies ΔE_{ij} (Li et al. 2011). In effect, this model expresses the energetics contents of TSP as a matrix of temperature-dependent transition probabilities describing the hopping evolution of a basis vector whose elements are the local energy minima. With this model in mind, the application of the Green–Kubo (linear response) theory of transport leads to an expression for $\eta(T)$ that one can explicitly evaluate. This was discussed in more detail in essay 10 for the fragile behavior analysis.

The same effective activation barrier analysis has been applied to supercooled liquid silica SiO_2 (Kushima et al. 2009b). Figures 11.1 and 11.2 show this is a contrasting case of *strong* temperature behavior. In the same spirit as the BLJ potential model for fragile liquids, a simple pair potential which is charge-independent was adopted (Feuston and Garofalini 1988). The effective activation barrier obtained is shown in figure 11.3. Compared to a typical barrier for *fragile* liquids (figure 10.8), the activation barrier is also seen to rise from a low value at high temperature to a high value at low temperature, although the rise is not as sharp. With additional analysis one can also extract from the TSP trajectory the detailed atomic configurations associated with the activation over specific barriers, and in this way obtain direct information on the activation mechanisms (Kushima et al. 2009a; Kushima et al. 2009b). Referring to the activation barriers marked α and β respectively in figure 11.3, one finds configurations associated with high barrier activation indicate a bond-switching process, whereas those associated with low-barrier activation to involve atomic rearrangements (rebonding) in the presence of dangling

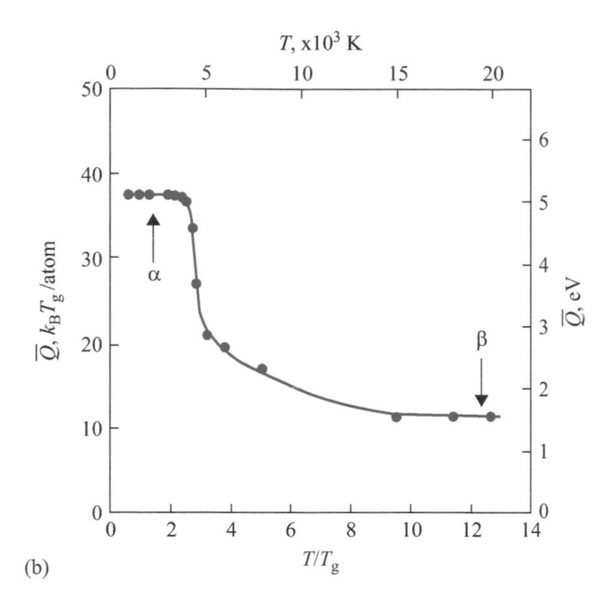

Figure 11.3
Variation of effective activation barrier with quench temperature obtained by the SPA model for supercooled liquid SiO_2. High and low activation barriers where atomic configurations associated with activation processes have been examined are indicated by α and β respectively (see text). Reprinted from (Kushima et al. 2009b), with the permission of AIP Publishing.

bonds (Kushima et al. 2009b). These mechanisms are quite distinct from those observed in the activated processes in a fragile system (Kushima et al. 2009a).

Figure 11.4 shows the viscosity results when figure 11.3 is inserted into equation (11.1). The predicted $\eta(T)$ displays a kink-like decrease at low temperature, beyond the range where measurements have been reported. In the low-temperature region, three viscosity results on SiO_2, each obtained by a different calculation have been reported. All are shown in figure 11.4 in absolute viscosity units and the temperature in K. First, we have the calculations using the Green–Kubo formula and MD simulations, which are reliable but exist only in the low-viscosity region (Kushima et al. 2009b; Horbach and Kob 1999). Then we have experimental measurements that exist only in the intermediate- to high-viscosity range (Angell 1988). Lastly, we have results of the present method extending over the entire range of values and overlapping with both the MD results and the experiments where they are available (Kushima et al. 2009b). Looking at the low-viscosity region, below $\eta \sim 1$ $Pa \cdot s$, two sets of MD results are shown, both computed using the Green–Kubo formula with MD simulations. The two lowest data points (open circles) are calculations (Kob and Andersen 1995) using a nonequilibrium MD (NEMD)

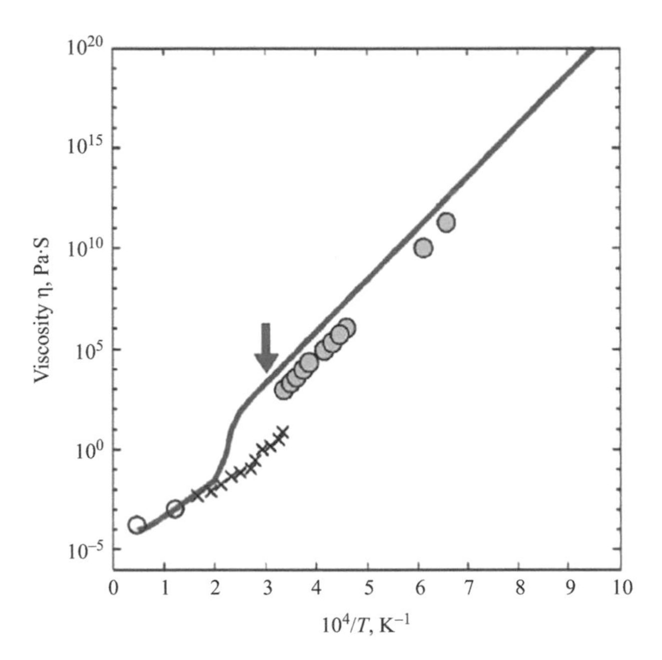

Figure 11.4
Viscosity of SiO_2, present calculation using the SPA model and the activation barrier (solid line) (Kushima et al. 2009a), experimental data (closed circles) (Angell 1988), and molecular simulation results via NEMD (open circles) (Kushima et al. 2009b), and MD (crosses) (Horbach and Kob 1999). The arrow indicates a crossover between fragile and strong behavior suggested by considerations of entropy and diffusivity effects in a model SiO_2 (Saika-Voivod, Poole, and Sciortino 2001). Reprinted from (Kushima et al. 2009b), with the permission of AIP Publishing.

method. The value of the second NEMD viscosity point is used to set the prefactor η_o in equation (11.1). The other eleven data points (crosses) are the results of a large-scale MD simulation (Horbach and Kob 1999) using a more sophisticated SiO_2 potential model. These may be considered the most accurate MD results available on SiO_2.

In the low-viscosity region, we see a good match between NEMD and the first five MD points. This suggests the effects of different potentials (Feuston and Garofalini 1988; Horbach and Kob 1999) are not significant at very low viscosities, $\eta \leq 10^{-3}\ Pa \cdot s$. We also see that SPA results (Kushima et al. 2009b) based on activated state kinetics through $\bar{Q}(T)$, are consistent with the totally independent MD results out to the first five points (two circles and three crosses). We take this to be a measure of validation of the present approach, albeit in a limited temperature range. The next six viscosity points (all crosses) from MD simulation show a discernible change in slope relative to the first five points, occurring around 4,000 K (Horbach and Kob 1999). While the possible significance of this feature was not discussed originally, we may now interpret this as possibly a crossover from strong to fragile behavior, as suggested by the SPA results.

The most stringent test of the molecular calculations lies in the high-viscosity range, 10^2–10^{12} $Pa \cdot s$. In figure 11.4, we have a direct comparison between the SPA results and experiment. An Arrhenius temperature variation is well predicted, while a systematic overestimate of the viscosity magnitude is also indicated. The latter should not be surprising considering the nature of the SPA approximation noted previously and the fact a similar behavior was also found for *fragile* liquid (Stillinger 1988). From figure 11.4 we see that the glass transition temperature T_g, defined by $\eta(T_g) = 10^{12}$ $Pa \cdot s$, is predicted to be to be 1,580 K, whereas the experimental value is 1,446 K (Angell 1988). Again, the slightly higher T_g could be attributed to the single activation path approximation as an upper-bound estimate. From the experimental viscosity data, one finds the activation energy to be 5.33 eV (Urbain, Bottinga, and Rachet 1982). In our independent determination using TSP trajectory, the value we obtain is 5.27 eV, as can be seen in figure 11.3.

In the intermediate viscosity range, 10^{-4}–10^2 $Pa \cdot s$, the SPA results show a smooth transition between the portion that overlaps with the first five NEMD and MD results and the portion that spans the experimental data. This suggests an interpretation of two closely spaced crossovers. As the viscosity increases with decreasing temperature, one encounters a first kink, which we may regard as the high-temperature crossover, from *strong* to *fragile*, followed by a second kink, the low-temperature crossover, from *fragile* to *strong*. In this respect, the remaining six MD results could be an indication of a possible high-temperature crossover. On the basis of atomistic calculations of entropy and diffusion coefficient using the same potential as the MD simulations (Horbach and Kob 1999) a low-temperature crossover at 3,300 K has been proposed (Saika-Voivod, Poole, and Sciortino 2001). This is shown by the arrow in figure 11.4.

We take advantage of having results for a *strong* and a *fragile* liquids obtained in the same way to examine their corresponding PEL topography. Recall this was a topic previously addressed by Stillinger. Our results are also useful for pointing out the distinguishing features of the two types of temperature behavior. The topological characteristics of a multidimensional PEL can be explored through various measures. One approach is to map the configuration space into multiple minima and transition states (barriers or saddle points), and connecting the minima and saddles in the form of a "disconnectivity graphs" (Wales 2003; Becker and Karplus 1997). This is a way to describe the general shape and overall connectivity that define the system landscape. In figure 11.5, we compare the results for a fragile system and SiO_2. Each local minimum is indicated by the end point of a vertical line while a saddle point (transition state) is denoted by a vertex. The fragile system displays a multitude of splitting, strong fluctuations in depth of local

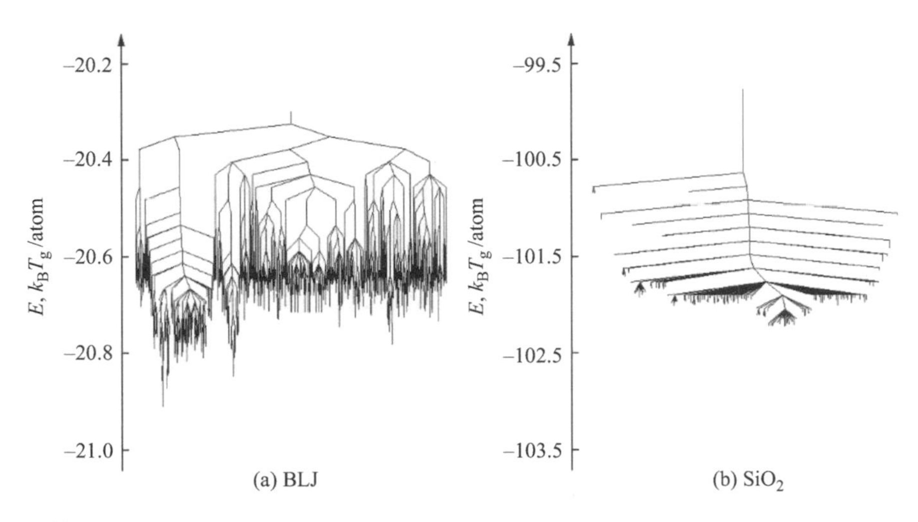

Figure 11.5
Disconnectivity graphs (Becker and Karplus 1997) of (a) BLJ and (b) SiO_2 constructed from their respective TSP trajectories. Reprinted from (Kushima et al. 2009b), with the permission of AIP Publishing.

minima, and significant basin connectivity, features that give the appearance of a "willow tree" in analogy with tree diagrams (Wales 2003). On the other hand, little of these features are evident in the graph for SiO_2. Generally speaking, PES structures may be classified as rough, single minimum, and funnel (Wales 2003). Each is associated with a distinctive disconnectivity graph, and a corresponding schematic of a one-dimensional cut of the 3*N*-dimensional PEL (Kushima et al. 2009b). The latter is particularly useful for visualizing and comparing different physical systems. The schematic potential profiles deduced from figure 11.5 are shown in figure 11.6. We see that the SiO_2 profile may be characterized as a broad-base funnel with relatively small fluctuations in the depth of local minima, with an overall "smooth" appearance. This is in contrast to the profile for BLJ, which indeed shows the features expected of a rough energy landscape (Stillinger 1988; Stillinger 1995; Debenedetti and Stillinger 2001; Kushima et al. 2009a). It is worth noting that the landscape profiles in figure 11.6 below are systematically deduced from quantitative data that are interatomic potential specific, from the TSP trajectory (figure 11.3) to the disconnectivity graphs (figure 11.5). Such comparisons provide a semi-quantitative way of relating system specifications at the level of interatomic potential and calculated activation barrier to a physical property, the viscosity.

We can now make a direct comparison of a fragile and a strong glass former, represented by model BLJ and SiO_2, respectively. This is shown in figure 11.7. Keep in mind that both types of supercooled liquids are treated consistently—molecular simulation

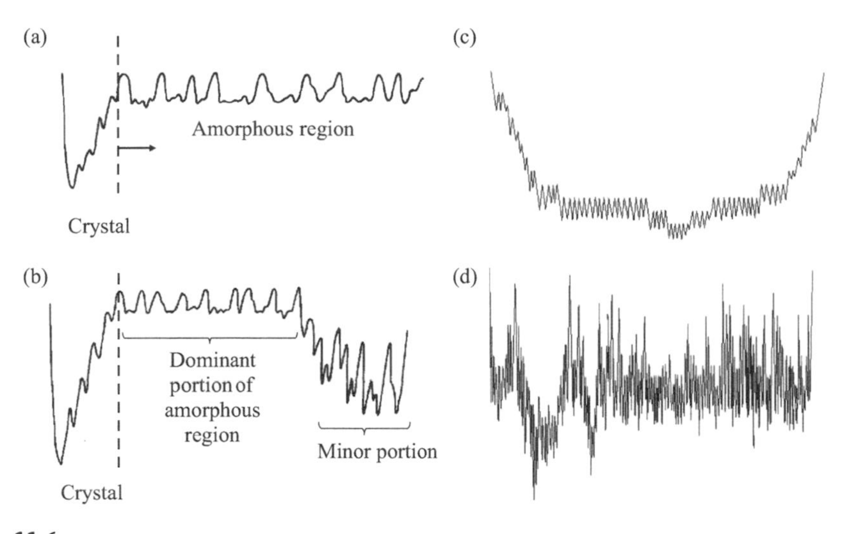

Figure 11.6
Schematic PEL profiles envisaged by Stillinger (Stillinger 1995; Debenedetti and Stillinger 2001) for (a) *strong* and (b) *fragile* liquids compared to one-dimensional cuts across the disconnectivity graphs in figure 11.5, profiles for (c) SiO_2 and (d) LBJ potential (Kushima et al. 2009b). Note the relative magnitude of fluctuations in barrier heights between a "smooth" and a "rough" energy landscape. Figures (a) and (b) reprinted from (Stillinger 1988), with the permission of AIP Publishing. Figures (c) and (d) reprinted from (Kushima et al. 2009b), with the permission of AIP Publishing.

using the ABC algorithm and data interpretation using TST. While at the viscosity level, the two systems could show quite distinct behavior—the basis of fragile versus strong characterization—at the effective barrier level, their basic behaviors have noteworthy similarities. Specifically, as temperature decreases, Q increases in both cases, albeit more steeply for the fragile system (inset of figure 11.7). From the standpoint of PES, this indicates the local minimum energy basin for the fragile system is deeper relative to the strong system, and this difference is sufficient to explain the markedly different viscosity behavior.

Toward a Unified View of Strong and Fragile Liquids

Following up on the similarity between the effective activation energies of strong and fragile viscous liquids shown in figure 11.7, one can notice both barriers have in common the basic structure sketched schematically in figure 11.8. This generic temperature-dependent activation barrier has two limiting values at low and high T, denoted as Q_H and Q_L respectively, and a smooth interpolation in the intermediate region demarcated by the

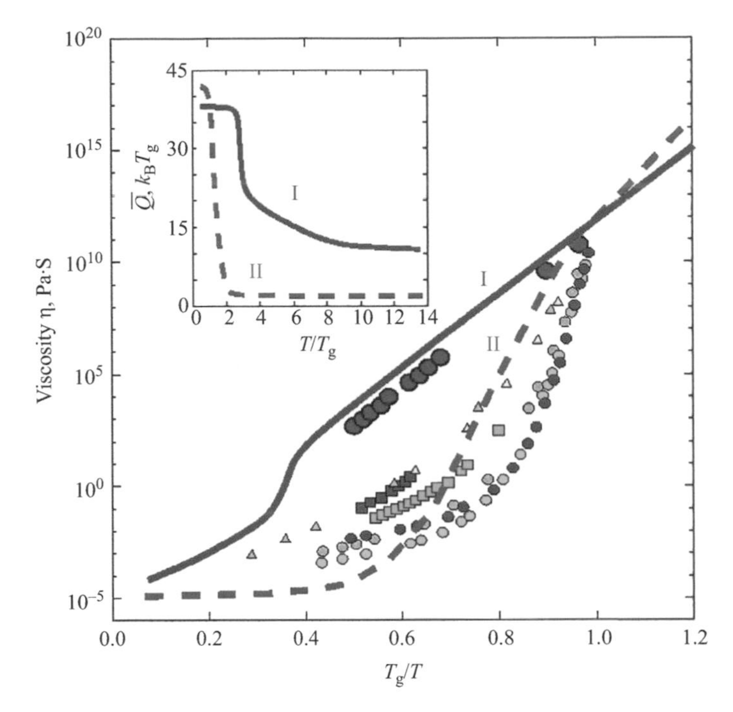

Figure 11.7
Comparison of shear viscosity of liquid silica, a strong glass former, calculated using the heuristic formulation (Kushima et al. 2009b) (solid curve I) with experimental data (closed circles) (Kushima et al. 2009b). Also shown are results for fragile liquids, calculation (Kushima et al. 2009a) (dashed curve II) and experimental data (various symbols) in the format of figure 11.2 (Angell 1988; Debenedetti and Stillinger 2001). Inset shows the corresponding $\bar{Q}(T)$ for the two supercooled liquid models in the same format as figure 11.3 and also figure 11.8. Reprinted from (Kushima et al. 2009b), with the permission of AIP Publishing.

temperatures T_L and T_H. The physical picture depicted is the following. When the system is evolving at temperatures above T_L, it encounters only shallow potential wells and therefore requires only a low activation energy Q_L. But if the system is evolving at temperatures below T_H, it is likely to be trapped in deep potential wells and therefore will require a high activation energy Q_H. The temperature variation of the generic barrier, $Q(T)$, therefore reflects an interpolation between the two limits. To determine $Q(T)$ qualitatively, it is sufficient to fix the set of four physical parameters (Q_H, Q_L, T_H, T_L). The transition range between T_L and T_H delineates the *fragility zone*; outside of this range $Q(T)$ is a constant. With figure 11.8 in mind, one can appreciate the commonality between a *strong* and a *fragile* liquid, and by extension their respective viscosities in figures 11.1 and 11.2. In the particular case of SiO_2 and BLJ, we see that the high activation barrier magnitudes are

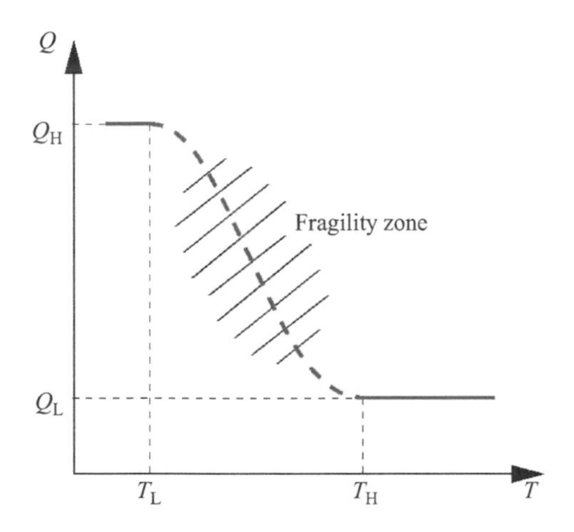

Figure 11.8
Schematic of a generic coarse-grained effective activation barrier with a bilevel structure (Kushima, Lin and Yip 2009). All labels are defined in the text. Reprinted from (Kushima et al. 2009b), with the permission of AIP Publishing.

similar, Q_H, ~ 40–50 $k_B T_g$, whereas the low activation barriers are quite different, Q_L ~ 10 $k_B T_g$ (SiO_2) versus 2 $k_B T_g$ (BLJ). Since Q_H (Q_L) governs the slope $\frac{d\eta}{dT}$ at T_H (T_L), this is the effect seen in the inset of figure 11.7. The large difference in Q_L also leads to an appreciable difference in the extent of the fragility zone, considerably smaller for SiO_2 than for BLJ, which in turn explains the pronounced fragile behavior of the latter. To explain why Q_L is so much larger for SiO_2 than BLJ, one can turn to an examination of the activation mechanisms in the respective studies (Kushima et al. 2009a, Kushima et al. 2009b).

A direct consequence of figure 11.8, self-evident by inspection, is the existence of two transitions (crossovers) between *strong* and *fragile* behavior. A system starting out at temperature above T_H should show Arrhenius viscosity variation. As T decreases below T_H, a crossover to *fragile* behavior should occur. With further temperature decrease the system maintains its *fragile* behavior until T gets below T_L, at which point a second crossover to *strong* behavior should set in. Thus, all glass formers should be expected to exhibit two crossovers (Kushima et al. 2009b).

Discussion

The intent of this essay is to examine the relationship between the temperature behavior of viscous flow of supercooled liquids and the understanding of the glass transition. The GAS challenge essentially asks for a molecular explanation, based on the notion of potential-energy landscape, of the observed viscosity variations shown in figure 11.2.

Here, we offer two such explanations. Using a sampling method for generating TSP trajectories (see figure 11.4), we constructed the effective activation barrier incorporating the combined role of the energetics (thermodynamics) of supercooled liquids and the kinetics of energy-barrier activation, in the spirit set forth by Goldstein (1969). The comparison of $\bar{Q}(T)$ for a representative *fragile* supercooled liquid and supercooled liquid SiO_2 suggests that the essential experimental behavior highlighted in figure 11.2 may have a simple, common origin as expressed in figure 11.8.

Regarding molecular calculations of $\eta(,)$ for direct confrontation with experimental data we find that while a single TSP trajectory may lead to temperature variations qualitatively similar to experiments for *fragile* supercooled liquids (figure 11.8), there are significant aspects of the experimental data over a range of some ten orders of magnitude that cannot be properly described without entropy corrections (Kushima et al. 2011), or without using several TSP trajectories in the network model calculation. With no fitting parameter used in all the calculations discussed, the present approach demonstrates a molecular-level predictive capability called for by the three preeminent investigators we had singled out.

The present work focuses only on a particular study of supercooled liquids and implications for the glass transition. It would be appropriate to note that many other relevant aspects of the phenomenon have been addressed in the statistical mechanics literature. Several recent monographs (Debenedetti 1996; Götze 2007; Binder and Kob 2011; Das 2011; Argon 2015), reviews (Angell 1995; Das 2004; Dyer 2006), and a special issue (Falk, Egami, and Sastry 2011) attest to the richness and fundamental interest of the glass transition.

In summarizing our response to the GAS challenge, we believe that the merits of the potential-energy landscape concept has been demonstrated (Yip 2016). The *fragile* versus *strong* behavior appears to have a common origin—activation and relaxation on the potential-energy landscape. These results suggest the same underlying mechanisms should operate generally in glass-forming liquids. Then the manifestation of *fragile* or *strong* behavior is governed by the specificity of the interatomic interaction (see figure 11.8). Additionally, the molecular mechanisms of viscous flow could be more universal than just the glass transition phenomenon. The concept of an effective activation barrier $Q(T)$ in TST gives rise to an isomorphism between $Q(T)$ and $Q(\sigma)$, where σ is an applied stress, thereby linking phenomena of thermal activation (viscosity) to stress-activated processes in rheology (Kushima et al. 2011). See also essays 13–15. This line of inquiry helps to frame an MSS in the emerging stage (Yip and Short 2013). The quintessential hallmark of slow relaxation in viscous flow that is our primary concern here thus becomes a significant illustration of a class of materials aging phenomena in which slow dynamics is a signature attribute. Well-known problems such as creep, stress corrosion cracking, cement setting, and even radiation damage may become amenable to simulation and analysis in the manner discussed here (Yip 2016; Yip and Short 2013; Short and Yip 2015).

References

Andrade, E. N. 1930. "The viscosity of liquids." *Nature* 125: 309.

Angell, C. A. 1988. "Perspective on the glass transition." *Journal of Physics and Chemistry Solids* 49: 863.

Angell, C. A. 1995. "Formation of glasses from liquids and biopolymers." *Science* 267: 1924.

Argon, A. S. 2015. *The Physics of Deformation and Fracture of Polymers*. Cambridge: Cambridge Univ. Press.

Becker, O. M., and M. Karplus. 1997. "The topology of multidimensional potential energy surfaces: theory and application to peptide structure and kinetics." *Journal of Chemistry and Physics* 106: 1495.

Binder, K., and W. Kob. 2011. *Glassy Materials and Disordered Solids*. Singapore: World Scientific.

Cao, P., M. Li, R. J. Heugle, H. S. Park, and X. Lin. 2012. "Self-learning metabasin escape algorithm for supercooled liquids." *Physical Review E* 86: 016710.

Das, S. P. 2004. "Mode-coupling theory and the glass transition in supercooled liquids." *Reviews of Modern Physics* 76: 785.

Das, S. P. 2011. *Statistical Physics of Liquids at Freezing and Beyond*. Cambridge: Cambridge Univ. Press.

Debenedetti, P. G. 1996. *Metastable Liquids, Concepts and Principles*. Princeton: Princeton Univ. Press.

Debenedetti, P. G., and F. H. Stillinger. 2011. "Supercooled liquids and the glass transition." *Nature* 410: 259.

Dyre, J. C. 2006. "Colloquium: the glass transition and elastic models of glass-forming liquids." *Reviews of Modern Physics* 78: 953.

Eyring, H. 1935. "The activated complex in chemical reactions." *Journal of Chemistry and Physics* 3: 107.

Falk, M., T. Egami, and S. Sastry. 2011. "Glass physics: from fundamentals to applications." *European Physics Journal E* 34: 85.

Fan, Y., S. Yip, and B. Yildz. 2014. "Autonomous basin climbing method with sampling of multiple transition pathways: application to anisotropic diffusion of point defects in hcp Zr." *Journal of Physics: Condensed Matter* 26: 365402.

Fan, Y., Y. N. Osetskiy, S. Yip, and B. Yildiz. 2013. "Mapping strain rate dependence of dislocation-defect interactions by atomistic simulations." *Proceedings of the National Academy of Science* 110: 17756.

Feuston, B. P., and S. H. Garofalini. 1988. "Empirical three-body potential for vitreous silica." *Journal of Chemistry and Physics* 89: 5818.

Goldstein, M. 1969. "Viscous liquids and the glass transition: a potential energy barrier picture." *Journal of Chemistry and Physics* 51: 3728.

Götze, W. 2007. *Complex Dynamics of Glass Forming Liquids*. Oxford: Oxford Univ. Press.

Horbach, J., and W. Kob. 1999. "Static and dynamic properties of a viscous silica melt." *Physical Review B* 60: 3169.

Kob, W., and H. C. Andersen. 1995. "Testing mode-coupling theory for a supercooled binary Lennard-Jones mixture I: the van Hove correlation function." *Physical Review E* 51: 4626.

Kushima, A., J. Eapen, J. Li, S. Yip, and T. Zhu. 2011. "Time scale bridging in atomistic simulation of slow dynamics: viscous relaxation and defect activation." *European Physics Journal B* 82: 271.

Kushima, A., X. Lin, J. Li, J. Eapen, J. C. Mauro, X. Qian, P. Diep, and S. Yip. 2009a. "Computing the viscosity of supercooled liquids." *Journal of Chemistry and Physics* 130 (22): 224504.

Kushima, A., X. Lin, J. Li, X. Qian, J. Eapen, J. C. Mauro, P. Diep, and S. Yip. 2009b. "Computing the viscosity of supercooled liquids. II. Silica and strong-fragile crossover behavior." *Journal of Chemistry and Physics* 131: 164505.

Kushima, A., X. Lin, and S. Yip. 2009. "Commentary on the temperature-dependent viscosity of supercooled liquids: a unified activation scenario." *Journal of Physics: Condensed Matter* 21: 504104.

Laidler, K. J., and M. C. King. 1983. "The development of transition-state theory." *Journal of Physics and Chemistry* 87: 2657.

Li, J., A. Kushima, J. Eapen, X. Lin, X. Qian, J. C. Mauro, P. Diep, and S. Yip. 2011. "Computing the viscosity of supercooled liquids: Markov network model." *PLos ONE* 6(3): e17909.

Martin, P. C. 1968. *Measurements and Time Correlation Functions*. New York: Gordon Breach.

Mazenko, G. F., and S. Yip. 1977. "Renormalized kinetic theory of dense fluids." In *Statistical Mechanics, Part B: Time-Dependent Processes*, edited by B. J. Berne, chap. 3. New York: Plenum Press.

Saika-Voivod, S., P. H. Poole, and F. Sciortino. 2001. "Fragile-to-strong transition and polyamorphism in the energy landscape of liquid silica." *Nature* 412: 514.

Short, M. P., and S. Yip. 2015. "Materials aging at the mesoscale: kinetics of thermal, stress, radiation activations." *Current Opinion in Solid State and Molecular Science* 9: 245.

Stillinger, F. H. 1988. "Supercooled liquids, glass transitions, and the Kauzmann paradox." *Journal of Chemistry and Physics* 88: 7818.

Stillinger, F. H. 1995. "A topographic view of supercooled liquids and glass formation." *Science* 267: 1935.

Urbain, G., Y. Bottinga, and P. Richet. 1982. "Viscosity of liquid silica, silicates and alumino-silicates." *Geochimica et Cosmochimica Acta* 46: 1061.

Wales, D. J. 2003. *Energy Landscapes*. Cambridge: Cambridge University Press.

Wigner, E. 1932. "On the quantum correction for thermodynamic equilibrium." *Physical Review* 40: 749.

Yip, S. 2016. "Understanding the viscosity of supercooled liquids and the glass transition through molecular simulation." *Molecular Simulation* 42: 1330.

Yip, S., and M. P. Short. 2013. "Multiscale materials modelling at the mesoscale." *Nature Materials* 12: 774.

Further Reading

Deng, D., A. S. Argon, and S. Yip. 1989. "A molecular dynamics model of melting and glass transition in an idealized two-dimensional material—I." *Philosophical Transactions of the Royal Society of London A* 329: 549.

Deng, D., A. S. Argon, and S. Yip. 1989. "Topological features of structural relaxation in a two-dimensional model atomic glass—II." *Philosophical Transactions of the Royal Society of London A* 329: 575.

Deng, D., A. S. Argon, and S. Yip. 1989. "Kinetics of structural relaxations in a two-dimensional model atomic glass—III." *Philosophical Transactions of the Royal Society of London A* 329: 595.

Deng, D., A. S. Argon, and S. Yip. 1989. "Simulation of plastic deformation in a two-dimensional atomic glass by molecular dynamics—IV." *Philosophical Transactions of the Royal Society of London A* 329: 613.

Kob, W. 1999. "Computer simulation of supercooled liquids and glasses." *Journal of Physics: Condensed Matter* 11: R85. Comprehensive review of advantages and drawbacks of MD simulation of the dynamics of supercooled liquids with discussions of both fragile and strong glass formers as well as the validity of the mode-coupling theory in the context of an ideal glass transition (in the absence of barrier activation).

Laughlin, R. B., D. Pines, J. Schmalian, B. P. Stojković, and P. Wolynes. 2000. "The middle way." *Proceedings of the National Academy of Science* 97: 32. Mesoscopic organizations in various forms of matter point to limits of current understanding of fundamental organizing principles and the emergence of new research frontiers.

Lu, J., G. Ravichandran, and W. L. Johnson. 2003. "Deformation behavior of the $Zr_{41.2}Ti_{13.8}Cu_{12.5}Ni_{10}Be_{22.5}$ bulk metallic glass over a wide range of strain-rates and temperatures." *Acta Materialia* 51: 3429. Stress and viscosity data allow a mechanism map in strain rate and temperature showing regimes of Newton, non-Newton flows, and shear localization.

Biroli, G. 2007. "A new kind of phase transition?" *Nature Physics* 3: 222. Jamming in 3D binary mixture of glass-forming particles at high densities and slowly driven externally are identified as hallmarks of dynamical heterogeneities.

Xu, N. 2009. "Equivalence of glass transition and colloidal glass transition in the hard-sphere limit." *Physical Review Letters* 103: 245701. MD simulations reveal a universal feature of glass formation emerging at low pressures and temperatures where the relaxation time can be scaled onto a single master plot as a single variable T/p. Thus, the colloidal glass transition at low pressure and the molecular glass transition at low temperatures are manifestations of the same phenomenon.

Mallemace, F., C. Branca, C. Corsaro, N. Leone, J. Spooren, S.-H. Chen, and H. E. Stanley. 2010. "Transport properties of glass forming liquids suggest that dynamical crossover temperature is as important as the glass transition temperature." *Proceedings of the National Academy of Science* 107: 22457. Experimental data on shear viscosity and self-diffusion show a characteristic crossover temperature T_x from Arrhenius to super-Arrhenius behavior, below which the Stokes–Einstein relation breaks down.

Bouchbinder, E., and J. S. Langer. 2011. "Linear response theory for hard and soft glassy materials." *Physical Review Letters* 106: 148301. Explanation of similar linear rheological properties of structural, metallic, and colloidal glasses based on shear transformation zone theory extended to include a broad distribution of internal thermal activation barriers found experimentally.

Beancourt, B. A. P., J. F. Douglas, and F. W. Starr. 2014. "String model for the dynamics of glass-forming liquids." *Journal of Chemistry and Physics* 140: 204509. Characteristics of string-like particle rearrangement deduced from simulation of dense-polymer melt of "spring-bead" model.

Langer, J. S. 2014. "Theories of glass formation and the glass transition." *Reports on Progress in Physics* 77: 042601. Commentary on various theoretical models on the dynamics of glass forming liquids.

Krausser, J., K. Samwer, and A. Zaccone. 2015. "Interatomic repulsion softness directly controls the fragility of supercooled metallic melts." *Proceedings of the National Academy of Science* 112 (45): 13762–13767. Probes temperature dependence of shear modulus and fragility behavior, comparing with results on colloidal glass.

Luo, P., Y. Zhai, P. Falus, V. García Sakai, M. Hartl, M. Kofu, K. Nakajima, A. Faraone, & Y Z. 2022. "Q-dependent collective relaxation dynamics of glass-forming liquid $Ca_{0.4}K_{0.6}(NO_3)_{1.4}$ investigated by wide-angle neutron spin-echo." *Nature Communications* 13: 2092. Spatial correlation effects showing change in relaxation mechanisms around 2.6 A, as indications of dynamical heterogeneity.

Tung, C. H., S.-Y. Chang, S. Yip, Y. Wang, J.-M. Carrillo, B. G. Sampter, Y. Shinohara, C. Do, W.-R. Chen. 2023. "Heterogeneities in viscoelastic relaxation and topological fluctuations in moderately supercooled liquids." preprint.

Scalliet, C., B. Giuslin, and L. Berthier. 2022. "Thirty milliseconds in the life of a supercooled liquid." *Physical Review X* 12: 041028. Relaxation dynamics of model supercooled liquids simulated by molecular dynamics over a time range of 10 orders of magnitude.

12 Strain-Rate Effects

Strain-rate effects are intrinsic to all materials deformations, in theory as well as modeling, experiments, or molecular simulations. When the deformation rate is significantly different from the rate of observation or analysis, it is unlikely that the underlying mechanisms governing the deformation at very different rates can be properly revealed or inferred. We investigate a feedback-localization phenomenon in the propagation of a single dislocation in a crystalline metal over a range of strain rates. The upturn behavior of the flow stress suggests a crossover from thermal activation to strain localization. By further considering the process of dislocation interaction with an intrinsic defect such as a self-interstitial atom cluster, one can construct a mechanism map in temperature and strain rate. Lastly, we comment on an analogy between crystal plasticity and glass rheology where strain rate plays a fundamental and structure-aware role.

Introduction

At low temperature, the deformation of metals is largely governed by the thermal activation of dislocation glide (Hoge and Mukherjee 1977). Experiments on different structures of metals indicate dislocation flow stress varies with strain rate in an apparently universal manner (Fan and Cao 2020). The flow stress slowly increases in an Arrhenius manner at low strain rates but turns sharply upward beyond a certain characteristic value. Although the results for different metals can be quantitatively different, the flow stress "upturn" behavior when the strain rate reaches the range of 10^3–10^4 s^{-1} (Regazzoni, Kocks, and Follansbee 1987; Armstrong, Arnold, and Zerilli 2009) appears to signify a fundamental origin. The onset of non-Arrhenius response has motivated several empirical constitutive models (Remington et al. 2006) to include phonon-drag effects in order to account for the data at high strain rates. All existing models to date invoke adjustable parameters to bridge the flow stresses below and above the critical strain rate.

Single Dislocation Propagation: Transition State Theory Modeling

We have previously encountered the concept of a Peierls barrier in essay 8, and PEL in essays 10 and 11. Here, we consider a dislocation line located initially on the bottom of the Peierls energy valley until a thermal fluctuation enables it to climb over the activation barrier and glide to the next valley (see figure 12.1). When an external stress is applied, the activation barrier $E(\sigma)$ is decreased, making it easier for the dislocation to escape from the valley. Based on this scenario, the escape rate is given by the TST

$$k(\sigma) = k_0 e^{-[E(\sigma)/k_B T]} \tag{12.1}$$

where k_0 is the attempt frequency. The activation barrier $E(\sigma)$ can be either empirically postulated or extracted from simulation carried out at a prescribed stress (Fan et al. 2012)

In the elastic deformation regime, the variation of stress with applied strain rate can be physically expressed as

$$\sigma = G\varepsilon = G\dot{\varepsilon}t \tag{12.2}$$

where G is the shear modulus and ε is the elastic strain because we are considering the initiation of dislocation flow at the onset of plasticity. In view of equation (12.2), $k(\sigma)$ becomes time-dependent.

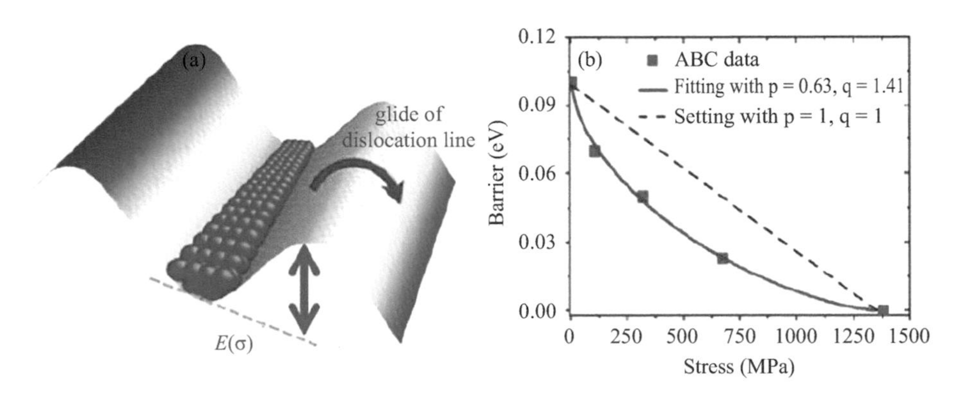

Figure 12.1
(a) Schematic of a single dislocation moving across a Peierls energy surface. (b) Activation barrier for dislocation glide decreases with increasing applied shear stress. The squares denote values extracted from simulation using the ABC method described in essay 10, and the solid curve is a fit to the stress-dependent empirical expression $E(\sigma) = E_0[1 - (\sigma/\sigma_c)^p]^q$ which is also mentioned in the inset in the text. The dashed line represents a hypothetical linearly decreasing profile. Reprinted by permission from Springer Nature: (Fan and Cao 2020).

To proceed with the analysis of variation of flow stress with strain rate, we introduce the residence probability $P(t)$ that the dislocation does not escape to a neighboring potential energy valley during time t. $P(t)$, in other words, is the probability that the system remains in the elastic deformation regime (Zhu et al. 2008), and it can be described by

$$\frac{dP(t)}{dt} = -k(t)P(t) \tag{12.3}$$

or

$$P(t) = \frac{1}{C}\exp\left[-\int_0^t k(t')dt'\right] \tag{12.4}$$

where C is the normalization factor. Consequently, the first-escape probability distribution $P(t)$ can be defined as

$$p(t) = \frac{dP(t)}{dt} = \frac{1}{C}k(t)\exp\left[-\int_0^t k(t')dt'\right] \tag{12.5}$$

with normalization

$$\int_0^{t_c} p(t)dt = 1 \Rightarrow C = \int_0^{t_c} k(t)\exp\left[-\int_0^t k(t')dt'\right]dt \tag{12.6}$$

where $t_c = \sigma_c/G\dot{\varepsilon}$ is the maximum residence time at a given strain rate $\dot{\varepsilon}$. The average residence time is then

$$\bar{t} = \int_0^{t_c} tp(t)dt = \frac{\int_0^{t_c} tk(t)\exp\left[-\int_0^t k(t')dt'\right]dt}{\int_0^{t_c} k(t)\exp\left[-\int_0^t k(t')dt'\right]dt} \tag{12.7}$$

In the limit of vanishing $\dot{\varepsilon}$, $k(t)$ is constant, and $t_c \to \infty$. Equation (12.7) shows the average time of residence as $\bar{t} = 1/k$, which is seen to follow an Arrhenius behavior. However, in general, at finite strain rate, $\bar{t}$ is expected to be non-Arrhenius.

It should be noted that the arguments based on equations (12.1)–(12.7) are applicable to any problem whenever the reaction rate is time-dependent. In the case of plastic flow at finite temperature and strain rate, the dislocation is expected to start gliding after a waiting period $\bar{t}$. Combining equation (12.2) with equation (12.7), one obtains

$$\bar{\sigma}_{flow} = G\dot{\varepsilon}\bar{t} = \frac{\int_0^{\sigma_c} \sigma k(\sigma)\exp\left[-\frac{1}{G\dot{\varepsilon}}\int_0^{\sigma} k(\sigma')d\sigma'\right]d\sigma}{\int_0^{\sigma_c} k(\sigma)\exp\left[-\frac{1}{G\dot{\varepsilon}}\int_0^{\sigma} k(\sigma')d\sigma'\right]d\sigma} \tag{12.8}$$

In summary, once the dislocation migration barrier profile $E(\sigma)$ is known, the flow stress for a given slip system under arbitrary thermomechanical conditions can be calculated according to equation (12.8), with only one parameter, the attempt frequency k_0 to be fixed. In practice, the value of k_0 will be of the order of the vibrational frequency of a typical solid, $\sim 10^{12}$ s^{-1}.

The deformation of bcc metals at low temperature is known to be controlled by the motion of ⟨111⟩/2 screw dislocations (Domain and Monnet 2005), the flow mechanism being 3D kink nucleation and propagation (Gordon et al. 2010; Rodney and Proville 2009). For testing equation (12.8), we consider a short dislocation segment of five Burger's vector, which should glide without kink nucleation (Fan et al. 2012). Returning to figure 12.1, the panel figure 12.1(b) shows the glide barrier for such a dislocation in bcc Fe. One sees a monotonically decreasing behavior well described by an expression of the form, $E(\sigma) = E_0[1 - (\sigma/\sigma_c)^p]^q$ (Rodney and Proville 2009; Kocks, Argon, and Ashby 1975), where $p = 0.63$, and $q = 1.41$. For purpose of benchmarking, we also include a dashed line in figure 12.1(b) with $p = q = 1$, which corresponds to the assumption of a linear behavior with a constant activation volume.

Figure 12.2 shows the thermal behavior of flow stress for strain rates varying over ten orders of magnitude, from $10^7\ s^{-1}$ to $10^{-3}\ s^{-1}$. In the low temperature limit, absent thermal activation, all flow stresses approach the Peierls stress, which is 1,400 MPa for the chosen interatomic potential. As temperature increases, the flow stresses monotonically decrease. At a fixed temperature, higher strain rate loading results in higher flow stress response, suggesting any quantitative comparison between experimental data and MD simulations must take into account the strain-rate differential.

The symbols in figure 12.2(a) represent the corresponding MD results at strain rate of $10^6\ s^{-1}$ and $10^7\ s^{-1}$. They are in reasonable agreement with the predictions of equation (12.8) using $k_0 = 1.2 \times 10^{12}\ s^{-1}$, derived from the Debye temperature. This constitutes a self-consistency test of equation (12.8) with $E(\sigma)$ taken from figure 12.1(b) (solid curve) in the strain-rate range of where MD is valid. One can see a sharp decrease of flow stress as the strain rate decreases to the range appropriate to conventional experiments. This variation is known experimentally to be a significant feature of thermal activation processes. It is not well captured by MD simulations at its characteristically high strain rates (Domain and Monnet 2005; Zhu et al. 2008). On the other hand, it is seen this behavior is qualitatively accounted for by the ABC simulation at the lower strain rates.

The variation of flow stress with strain rate is of fundamental interest in experimental studies of crystal plasticity. Figure 12.2(b) shows the predicted behavior based on equation (12.8). Under the limit of infinitely high strain rate, the flow stress approaches the Peierls stress. On the other hand, the flow stress is negatively sensitive to the

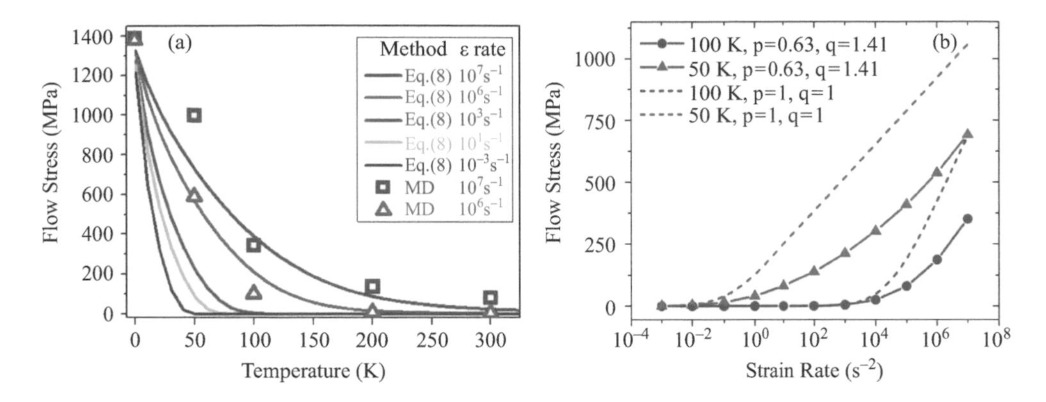

Figure 12.2
(a) Flow stress of the (111) screw dislocation in bcc Fe under different strain rate and temperature conditions. The solid lines show the calculated results by equation (12.8), and the data points are MD results at high strain rates. In the inset "Eq. (8)" refers to equation (12.8). (b) Predicted flow stress variation at different strain rates. Solid and dashed curves are calculated based on realistic (p, q) barrier profile and hypothetical linear profile shown in figure 12.1(b), respectively. Reprinted figure with permission from (Fan et al. 2012). Copyright 2012 by American Physical Society.

temperature. In the high temperature limit, the flow stress approaches zero regardless of the strain rate. At low $\dot{\varepsilon}$, the flow stress increases only moderately, but as $\dot{\varepsilon}$ increases, above 100 s^{-1} at 50 K, and $10^4\ s^{-1}$ at 100 K, it begins to increase much more strongly. This upturn behavior can be analyzed in terms of two factors: stress-dependent activation volume and strain rate–induced non-Arrhenius behavior. Because of the nonlinear stress dependence of the activation barrier (figure 12.1b), the activation volume is very small at high stresses. Such small activation volume leads to a high sensitivity of the flow stress dependence on strain rate (Zhu et al. 2008). In addition, as derived in equation (12.7), there is a non-Arrhenius behavior due to the strain rate loading that also contributes to the upturn in figure 12.2(b). To decouple the two contributions, we remove the nonlinearity of $E(\sigma)$ by setting p and q equal to unity (dashed line in figure 12.1b). Now the only nonlinear factor comes from the strain rate–induced non-Arrhenius behavior in equation (12.8). As shown in figure 12.2(b), under this condition, the flow stress upturn remains, but the stress is now higher beyond the crossover strain rate. Since the assumption of $p = q = 1$ results in a higher effective barrier and correspondingly a longer residence time, it follows that the flow stress response is higher as well. Our analysis therefore shows the onset of flow stress upturn is to be attributed mainly to the non-Arrhenius behavior induced by the strain rate, as described by equations (12.7) and (12.8).

Flow Stress Upturn

To compare the predicted upturn behavior quantitatively with experimental data, we adapt the energy profile $E(\sigma)$ for a longer screw dislocation system in bcc Fe (Gordon et al. 2010) and use it as the input into equation (12.8). Figure 12.3 shows the variation of flow stress and strain rate at 300 K, as observed experimentally and predicted by our model. Since the flow stress magnitude is strongly influenced by the defect microstructures in the experimental specimens (Armstrong, Arnold, and Zerilli 2009), the quantitative comparison can only be meaningful after an appropriate normalization. Therefore, in figure 12.3 we show the reduced flow stress, defined as the ratio of flow stress to its value at the highest strain rate $10^7\ s^{-1}$, as a function of strain rate. It is seen that both the experiments and our calculation results show a significant flow stress upturn with the critical strain rate in the range of 10^4–$10^5\ s^{-1}$. The extent of the agreement suggests that equation (12.8) plus $E(\sigma)$ have essentially captured the mechanism for the flow stress upturn behavior. On the other hand, it is known that the flow stress magnitude depends on the local defect microstructure in the material. Experimental specimens have a complex defect microstructure leading to appreciably higher flow stresses as seen in

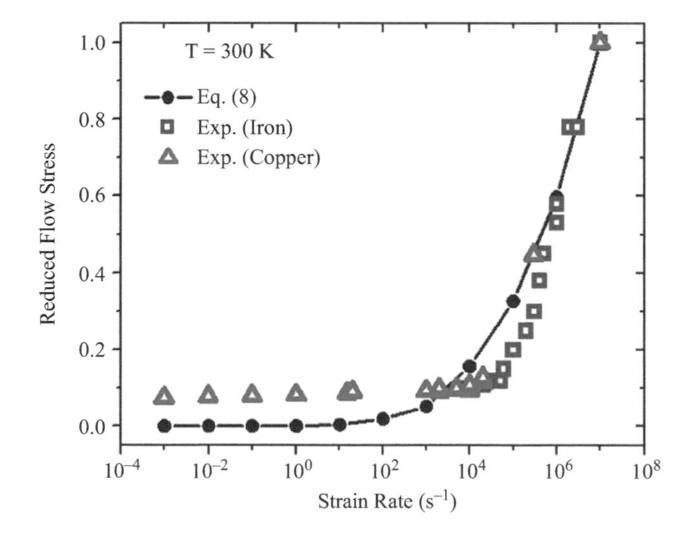

Figure 12.3
Variation of reduced flow stress with strain rate at 300 K. Results from equation (12.8) written as "Eq. (8)" in the inset are shown as closed circles fitted with a solid curve (Fan et al. 2012). Experimental data on Cu and Fe are adapted from (Armstrong, Arnold, and Zerilli 2009; Gordon et al. 2010). Reprinted figure with permission from (Fan et al. 2012). Copyright 2012 by American Physical Society.

the comparison in figure 12.2(b). It is therefore reassuring that the reduced flow stress predicted by our model in figure 12.3 is also quantitatively consistent with experiments on different materials, for instance ductile copper and brittle iron. This may be attributed to the fact the energy barrier for dislocation to move past existing defects in the material bears a similar stress-activated behavior as the simple dislocation glide represented by the expression $E(\sigma) = E_0[1 - (\sigma/\sigma_c)^p]^q$ previously discussed (Remington et al. 2006). Without invoking any specific mechanism or fitting parameters, a PEL-based approach can naturally explain the nonlinear transition of the flow stress from a classical Arrhenius behavior to a sharp upturn across a critical strain rate.

Probing the mechanisms of defect-defect interactions on the nanometer spatial scale and strain rates lower than $10^{-6}\,s^{-1}$ is a longstanding challenge for the MD-simulation community. An approach based on combining traditional TST with the concept of a strain-dependent effective activation barrier has been implemented to study the kinetics of dislocation-defect interaction over a wide range of strain rates (Fan et al. 2013). The defect investigated in this case is the self-interstitial atom cluster (SIA) in hcp-Zr. A novel strain rate–dependent trigger mechanism was found that allowed the SIA to be absorbed during the process, leading to dislocation climb. This finding was then used in a coarse grain-rate equation to construct the strain rate–temperature mechanism map.

Mechanism Map: Dislocation-SIA Interaction

Mechanism maps are useful for expressing how particular materials behavior can vary with two distinct state variables or dynamical descriptors (order parameters). We have previously encountered an example of this in essay 9 regarding the variation of interface strength with grain size. When the effects of two variables are coupled (or two mechanisms compete), regions where one variable dominates over the other, as well as limiting behavior, can be easily demarcated (crossover boundary). One can also imagine a stack of mechanism maps along a third variable. This way of summarizing what is known about the particular behavior in question is a general form of the fundamental paradigm of materials science widely known as *structure-property correlation.*

The mechanism map that one can construct for the present problem of dislocation-SIA interaction is shown in figure 12.4. The red curve is the predicted boundary separating two competing processes, dislocation by-passing the SIA (Recovery), and dislocation absorption of SIA and undergoing a jog (Climb). Blue squares denote MD results showing Recovery (open squares) and Climb outcome (closed squares), respectively. MS denotes molecular statics calculations (Griffiths 1993). The symbol exp. denotes experimental data (Onimus and Béchade 2009). Notice that the strain rate covered is $10^{-7} - 10^{7}\,s^{-1}$.

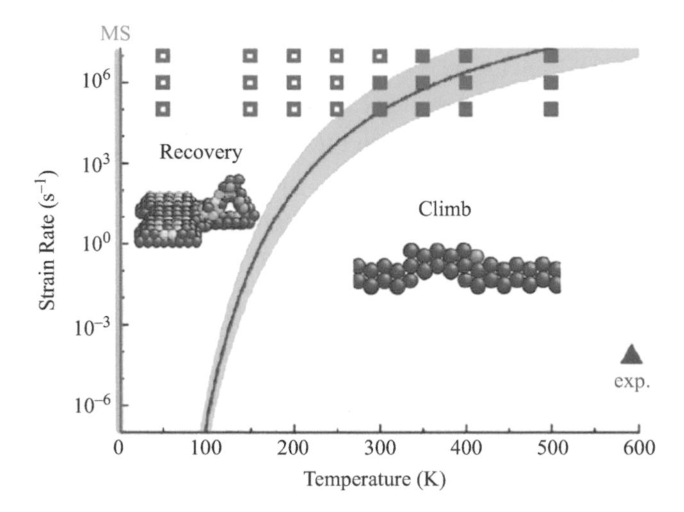

Figure 12.4
Mechanism map of dislocation-SIA cluster interaction in hcp-Zr showing regions where Climb processes predominate. Open and closed squares are direct MD simulations that are in agreement with the modeling predictions based on equation (12.8). Figure from (Fan et al. 2013).

Crystal Plasticity—Glass Rheology Analogy

The present study has been conducted by combining TST modeling with an extension of the metadynamics simulation algorithm ABC introduced in essay 10 and applied in essay 11. Here, we are focusing on the strain-rate effects on single dislocation propagation and the interaction between a dislocation-SIA cluster interaction. In both cases, competing interaction mechanisms involve an interplay between thermal activation and strain-rate effects, with different consequences on thermomechanical behavior. Such findings have broad implications for understanding material complexities through the integration of molecular simulation with theory and modeling and with experiments (Fan and Cao 2020).

Given that the present essay is concerned with crystal plasticity and essays 10 and 11 are concerned with glassy states of matter, one may wonder what commonalities exist between crystalline and glassy solids when it comes to the effects of strain rate on mechanical behavior. With considerable community interests in both types of solids, the question deserves attention from both sides (Fan, Yildiz, and Yip 2013; Fan and Cao 2020). In conclusion and by way of looking ahead to essay 14, we note that dynamic yielding at high strain rates is a universal phenomenon of strain localization. It is therefore an integral part of the challenge and opportunity of the mesoscale materials research frontier previously mentioned in the prologue and will be taken up in the epilogue.

References

Armstrong, R., W. Arnold, and F. J. Zerilli. 2009. "Dislocation mechanics of copper and iron in high rate deformation tests." *Journal of Applied Physics* 105 (2): 023511–023517.

Domain, C., and G. Monnet. 2005. "Simulation of screw dislocation motion in iron by molecular dynamics simulations." *Physical Review Letters* 95 (21): 215506.

Fan Y., and P. Cao. 2020. "Long Time-Scale Atomistic Modeling and Simulation of Deformation and Flow in Solids." In *Handbook of Materials Modeling*, edited by W. Andreoni and S. Yip. Switzerland AG: Springer Nature.

Fan, Y., B. Yildiz, and S. Yip. 2013. "Analogy between glass rheology and crystal plasticity: yielding at high strain rate." *Soft Matter* 9: 9511.

Fan, Y., Y. N. Osetsky, S. Yip, and B. Yildiz. 2012. "Onset mechanism of strain-rate-induced flow stress upturn." *Physical Review Letters* 109 (13): 135503.

Fan, Y., Y. N. Osetskiy, S. Yip, and B. Yildiz. 2013. "Mapping strain rate dependence of dislocation-defect interactions by atomistic simulations." *Proceedings of the National Academy of Science* 110 (44): 17756–17761.

Gordon, P. A., T. Neeraj, Y. Li, and J. Li. 2010. "Screw dislocation mobility in BCC metals: the role of the compact core on double-kink nucleation." *Modeling Simulation and Materials Science Engineering* 18 (8): 085008.

Griffiths, M. 1993. "Evolution of microstructure in hcp metals during irradiation." *Journal of Nuclear Materials* 205: 225–241.

Hoge, K. G., and A. K. Mukherjee. 1977. "The temperature and strain rate dependence of the flow stress of tantalum." *Journal of Materials Science* 12 (8): 1666–1672.

Kocks, U. F., A. S. Argon, and M. F. Ashby. 1975. "Thermal Activation." In *Progress in Materials Science*, edited by B. Chalmers, J. W. Christian, T. B. Massalski, 19 (9): 110. Oxford: Pergamon Press.

Onimus, F., and J.-L. Béchade. 2009. "A polycrystalline modeling of the mechanical behavior of neutron irradiated zirconium alloys." *Journal of Nuclear Materials* 384 (2): 163–174.

Regazzoni, G., U. F. Kocks, and P. S. Follansbee. 1987. "Dislocation kinetics at high strain rates." *Acta Metallurgica* 35 (12): 2865–2875.

Remington, B. A., P. Allen, E. M. Bringa, J. Hawreliak, D. Ho, K. T. Lorenz, H. Lorenzana, J. M. McNaney, M. A. Meyers, S. W. Pollaine, K. Rosolankova, B. Sadik, M. S. Schneider, D. Swift, J. Wark, and B. Yaakobi. 2006. "Material dynamics under extreme conditions of pressure and strain rate." *Science Technology* 22 (4): 474–488.

Rodney, D., and L. Proville. 2009. "Stress-dependent Peierls potential: influence on kink-pair activation." *Physical Review B* 79 (9): 094108.

Zhu, T., J. Li, A. Samanta, A. Leach, and K. Gall. 2008. "Temperature and strain-rate dependence of surface dislocation nucleation." *Physical Review Letters* 100 (2): 025502.

Further Reading

Fan, Y., A. Kushima, S. Yip, and B. Yildiz. 2011. "Mechanism of void nucleation and growth in bcc Fe: atomistic simulations at experimental time scales." *Physical Review Letters* 106: 125501. ABC simulation of void nucleation and growth in bcc Fe at experimental timescales.

Fan, Y., S. Yip, and B. Yildiz. 2014. "Autonomous basin climbing method with sampling of multiple transition pathways: application to anisotropic diffusion of point defects in hcp Zr." *Journal of Physics: Condensed Matter* 26: 365402. ABC simulation of diffusion anisotropy in hcp-Zr.

Scalliet, C., B. Giuslin, and L. Berthier. 2022. "Thirty milliseconds in the life of a supercooled liquid." *Physical Review X* 12: 041028. Relaxation dynamics of model supercooled liquids simulated by molecular dynamics over a time range of 10 orders of magnitude.

V Soft-Matter Rheology

13 Amorphous Creep Mechanisms

Molecular processes of creep in a metallic thin film are simulated at experimental timescales using an atomistic method based on metadynamics. Space-time evolutions of the deviatoric strains and nonaffine particle displacements are analyzed to bring out details of atomic-level deformation and flow processes underlying the characteristic stress and temperature dependence of amorphous creep. From the simulation results, spatially resolved on the nanoscale and temporally coarse-grained over 0.1 second intervals, we derive an explanation of the experimentally well-known stress variation of creep rate. The simulation data lead to a mechanism map delineating the regimes of diffusional creep at low stress and high temperature, and deformational creep at high stress and low temperature. These findings bring together two longstanding views on the mechanisms of amorphous plasticity. Atomic diffusion via free volume and stress-induced shear deformation are naturally coupled in the emerging concept being referred to as dynamical heterogeneities—the slowly driven, interaction-dominated fluctuations characteristics of systems out of equilibrium.

Introduction

Deformation and flow are fundamental rheological processes known in many materials. For molecular understanding of plastic behavior under stress (creep), a standing challenge is to explain how the constituent atoms rearrange themselves individually and collectively in response to local mechanical and thermal stresses. Such information is necessary for the formulation of mechanistic models.

We address the question of identifying the elementary processes of deformation and flow in amorphous creep through atomistic simulation. A well-known bottleneck is that the *relevant* temporal scales in the laboratory are well beyond the reach of traditional MD simulation. As an alternative, we apply a metadynamics formulation that allows transition state trajectories to be generated while retaining molecular spatial resolution (see essays 10 and 11). With this approach, tensile creep in an atomistic model of a metallic glass on the timescale of seconds has been studied (Cao, Short, and Yip 2017). By analyzing the spatial distributions of deviatoric strain and nonaffine particle displacement, one observes the effects of stress and temperature on the evolution of activated states. The processes of single-particle diffusion and shear deformation of particle clusters are found to be intimately coupled. This coupling is essential to interpreting the characteristic upturn behavior of stress effects on creep rate, which is experimentally widely known but not explained previously in terms of molecular mechanisms. The simulation results also support a mechanism map showing a regime of low stress and high temperature where diffusional creep dominates, and a high-stress regime governed by shear deformation creep. These findings suggest that the two prevailing models of amorphous plasticity are complementary in their focus on either the single-particle (diffusive) or the collective (deformation) degrees of freedom.

F. Spaepen was the first to consider the mechanisms associated with homogeneous and inhomogeneous flows (Spaepen 1977). In the former, all volume elements of the system contribute to the strain, whereas in the latter, strain production and evolution is localized in only a few thin shear bands. Particularly emphasized was the single-particle diffusional aspect of flow, with the local free volume considered as a mechanistic order parameter. In this view, local strain production and dissipation are assumed to be associated with individual atomic jumps. At the time, the free-volume concept itself was under discussion in connection with the nature of the glass transition in supercooled liquids.

Shortly thereafter, A. S. Argon proposed a plastic deformation model of metallic glasses based on the notion of activated shear transformations (Argon 1979). At low temperatures, the transformations are assumed to be diffusive in nature producing relatively small strains. At high temperatures (and also high stresses), the transformations are sufficiently numerous they readily give rise to strain localization. In contrast to the diffusional mechanism, this view emphasized the role of stress-induced deformation as a collective (rather than single-particle) response.

Subsequently, Falk and Langer introduced the term *shear transformation zone* (STZ) in interpreting the results of MD simulation of viscoplastic deformation of a model metallic glass (Falk and Langer 1998), with the intention of building on the two foregoing models. While this term is now widely adopted, at times it is not precisely clear what aspect of the Spaepen or Argon model one is referring to. Here, we emphasize the complementary aspects of the Spaepen and Argon formulations. For clarity we will refer to the mechanism proposed by Spaepen as FAD (free-volume atomic diffusion) and that described by Argon as STD (shear-transformation deformation). We reserve the term STZ only for the statistical mechanical theory of deformation and failure of amorphous solids that has been developed (Falk and Langer 2011). It should be noted that the commonalities between the Spaepen and Argon models and the term STZ have been discussed interchangeably (Schuh, Hufnagel, and Ramamurty 2007).

Atomic-Level Strain and Nonaffine Displacement

There are two ways of defining atomic-level strain, deviatoric strain $D_{\min}^2$ and local von Mises strain η_{Mises}, as measures of local plastic deformation. Imagine that a region surrounding an atom undergoes a strain deformation during a time interval δt. Let the deviatoric strain be defined as

$$D_{\min}^2(t, t+\delta t) = \sum_{i=1}^{n} [\mathbf{X}_i(t+\delta t) - \mathbf{X}_0(t+\delta t) - \mathbf{J} \times (\mathbf{X}_i(t) - \mathbf{X}_0(t))]^2 \tag{13.1}$$

where $\mathbf{X}(t)$ is the reference configuration at time t, $\mathbf{X}(t+\delta t)$ is the current configuration at time $t+\delta t$, and the index i runs over all atoms within the interaction cutoff relative to the reference position of atom 0. $\boldsymbol{J}$ is the affine deformation tensor that transforms a nearest neighbor separation, $\mathbf{X}_i(t) - \mathbf{X}_0(t)$, to what would be expected under an affine deformation. The deformation tensor $\mathbf{J}$ is determined by minimizing $D_{\min}^2$ with the minimum value being the atomic-level deviatoric strain. For each $\mathbf{J}$, a Green strain tensor $\boldsymbol{E}$ can be written as $\boldsymbol{E} = \frac{1}{2}(\mathbf{J} \cdot \mathbf{J}^T - \mathbf{I})$. The local strain invariant η_{Mises} is computed by

$$\eta_{\text{Mises}} = \sqrt{\eta_{xy}^2 + \frac{\eta_{xx}^2 + \eta_{yy}^2 - \eta_{xx}^2\eta_{yy}^2}{3}} \tag{13.2}$$

Note that η_{Mises} is directly derived from $\mathbf{J}$, while $D_{\min}^2$ is a measure of the nonaffine deformation. Both quantities are used in studies of local plastic deformation in amorphous materials (Lu, Ravichandran, and Johnson 2003; Spaepen 1977; Argon 1979).

The nonaffine displacement is also useful for tracking single atom dynamics in steady state creep. For a time interval $(t, t+\delta t)$, it is defined as

$$\delta u_\alpha(t, t+\delta t) = X_\alpha(t+\delta t) - F_{\alpha\beta} X_\beta(t) \tag{13.3}$$

where Greek indices α and β indicate the Cartesian components, and the deformation gradient F is related to system-level creep strain $\varepsilon_{\alpha\beta}$ by $F_{\alpha\beta} = 1 + \Delta\varepsilon_{\alpha\beta}(t, t+\delta t) = 1 + (\varepsilon_{\alpha\beta}(t+\delta t) - \varepsilon_{\alpha\beta}(t))$.

Creep Curves

Figure 13.1 shows time-evolution of the system strain determined by the metadynamics simulation algorithm described in essay 10. The creep curve is seen to display the classical behavior of three stages of strain evolution. The initial period of strain nucleation and distribution, consisting of a steep increase followed by a gradual approach to saturation, is known as primary or transient creep. The secondary stage of steady-state

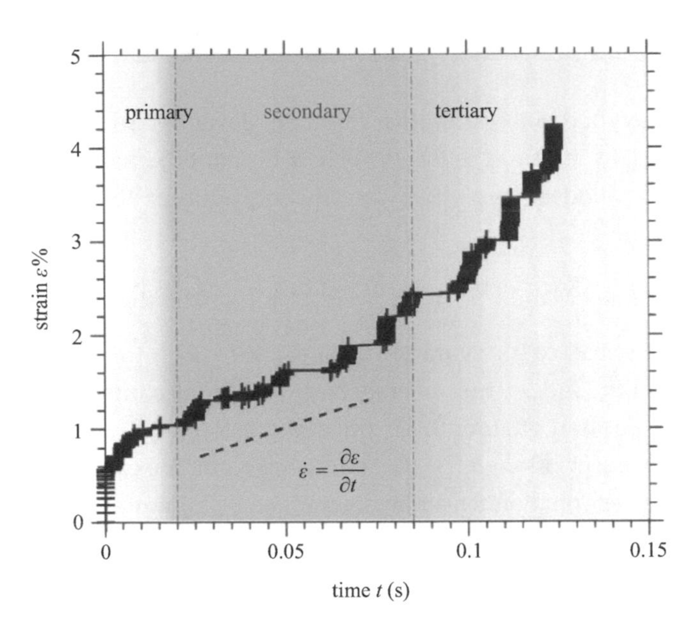

Figure 13.1
Time evolution of system strain of a thin film under tensile stress. Metadynamics simulation of a model metallic glass at 300 K (0.68 T_g) and stress of 198 MPa. Figure from (Cao, Short, and Yip 2017).

creep is a period of linear strain increase in time. The extent of this stage depends on the combination of applied stress and system temperature. For relatively low stresses, the secondary stage may have a considerable extent and then one may not reach the tertiary stage where strain rate undergoes a run-away increase without apparent limit. This would be an optimum situation from the standpoint of being able to follow all three stages of study of the creep mechanisms. This is the case with figure 13.1, as the simulated response shows the onset of structural instability at a strain level of ~4%. Notice that the relevant timescale of creep, as seen in the overall behavior of figure 13.1, is of the order of seconds. Recalling the characteristic timescales of MD simulations is of the order of picoseconds to nanoseconds, clearly it would not be feasible for MD to access the entire temporal range of figure 13.1, which is roughly comparable to those of laboratory measurements.

Assuming that the creep responses simulated at the system level are physically meaningful, then the corresponding atomic configurations along the evolution trajectory can be analyzed to provide the desired details of the individual atomic motion and collective rearrangements during the different stages of the creep. Such direct access to atomic-level details is the unique contribution that simulation can make in a synergistic study in combination with experiments and model calculations.

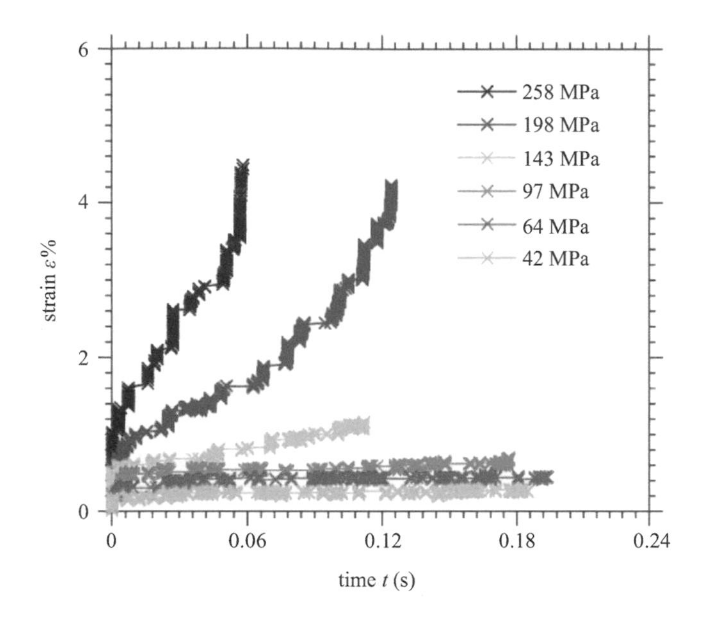

Figure 13.2
Simulated creep curves at indicated stress levels at temperature $T = 0.68T_g$. Figure from (Cao, Short, and Yip 2017).

Figure 13.2 shows a series of creep curves spanning the stress range from 42 to 258 MPa, all simulated at 300 K. At low stresses, one sees only the primary and secondary stages during the time interval of simulation. For these cases the creep rate, the slope in the secondary stage, is sufficiently small that the system remains in steady-state creep throughout the simulation. In contrast, for the two highest stresses one sees clearly the onset of tertiary creep, and correspondingly the extent of the secondary stage is progressively reduced. In this study, we focus on the mechanisms that sustain stage-2 creep; nevertheless, it is pertinent to note that tertiary creep will be triggered by the existence of a shear band with strain localization leading to structural instability (failure) of the material (Cao, Short, and Yip 2017).

Nonaffine Flow and Deviatoric Strain Distributions

The variation with stress at the system level, seen in figure 13.2, calls into question the underlying molecular rearrangements associated with local deformation and flow. This is the kind of information that can be deduced from the details of particle position, which atomistic simulation is uniquely capable of sampling. Figure 13.3 shows how the

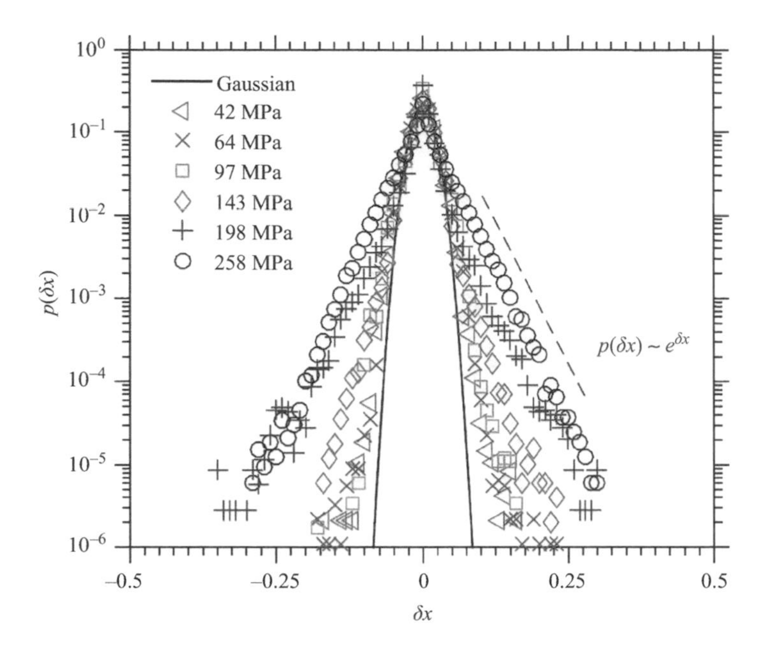

Figure 13.3
Statistical distributions of nonaffine particle displacement δx at $T = 300$ K (0.78 T_g) and the stresses indicated. Also shown are a Gaussian core (solid curve) and an exponential wing (dashed line). Figure from (Cao, Short, and Yip 2017).

probability distributions of nonaffine particle displacement δx vary with stress. Each distribution is seen to be composed of a Gaussian core and exponentially varying wings that become broader with increasing stress. This behavior has been observed in MD simulation of model glasses (Kob et al. 1997) and in experiments on colloidal particles (Weeks et al. 2000). They have been interpreted as localized fluctuations known as *dynamical heterogeneities*. The existence of significant fractions of particle moving appreciably faster or slower than the average is regarded a characteristic attribute of space-time fluctuations in driven systems. By inspection of figure 13.3, one can discern a change in the behavior of the probability distributions across 143 and 198 MPa. It indicates a gap in the stress variation of the statistical distributions. To see what this could mean in terms of molecular mechanisms, one could again turn to molecular details appropriately resolved spatially and temporally.

The nonaffine displacement δx may be regarded as the order parameter for diffusive atomic rearrangement. Similarly, the deviatoric strain D^2_{m}, could be treated as the order parameter for molecular rearrangement in deformation. Figure 13.4 shows the stress variation of the distributions D^2_{m}. The distributions indicate a power-law decay for small D^2_{m}

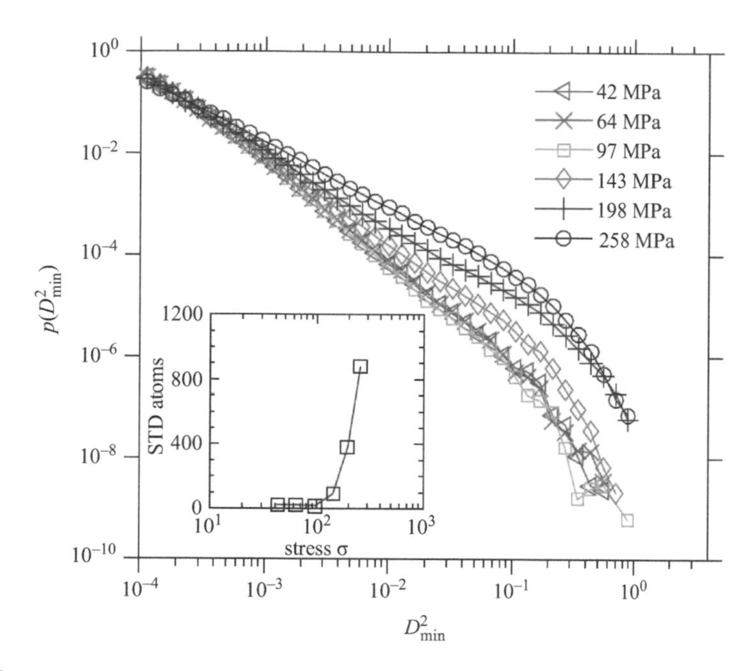

Figure 13.4
Statistical distributions of deviatoric strain D^2 at temperature $T = 300$ K (0.78 T_g) and the various stresses indicated. Also shown is the variation of number of atoms involved in shear transformation events with stress in MPa (inset). Figure from (Cao, Short, and Yip 2017).

with a characteristically slower drop-off at large D^2_m. Here, one also finds a gap between the distributions at 143 and 198 MPa. Counting the number of atoms participating in shear transformation events at each stress loading, we obtain the result shown in the inset of figure 13.4. Corresponding to the gap noted in figures 13.3 and 13.4, a sharp increase in the number of particles involved is seen. Thus, beyond a certain stress significantly more particles are participating in flow as well as in deformation. This is direct evidence that stress plays a fundamental role in the physical manifestation of creep responses. In other words, mechanical (external) deformation and internal flow are mutually sensitive to each other.

The local space-time distributions that give rise to the probability distributions discussed in figures 13.3 and 13.4 are presented in figure 13.5. Spatial maps of two types of creep measures are shown together. They are the results of analysis performed on the atomic trajectories produced by the simulations. The maps are spatially resolved at the nanoscale, and temporally measured over a time interval of order 0.1 second. Spatial maps of nonaffine displacements δx (local flows as opposed to homogeneous

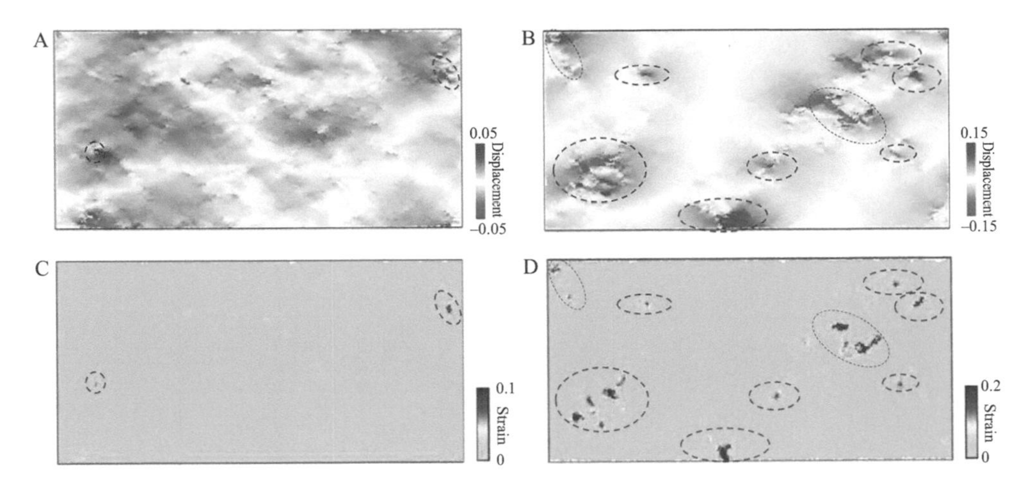

Figure 13.5
Spatial distributions of nonaffine displacement (a) and (b) and deviatoric strain (c) and (d) at temperature $T = 300$ K (0.78 T_g). At each level, left and right panels denote low stress (64 MPa) and high stress (258 MPa), respectively. All measures are taken over an interval of 0.012 seconds. Figure from (Cao, Short, and Yip 2017).

or system-wide responses) are shown as panels figure 13.5(a) and figure 13.5(b). Correspondingly, maps of deviatoric strain D^2_{m}, (local deformation) are seen in panels figure 13.5(c) and figure 13.5(d). For each set, the left-right panels are arranged as low (64 MPa) and high (258 MPa) stresses, respectively.

The layout of the panels is intended to facilitate the discussion of stress effects on flow and deformation in creep, as well as quantifying the intrinsic coupling between the processes at the molecular level. For the former, we compare figures 13.5(a) with 13.5(b), and figures 13.5(c) with 13.5(d), whereas for the latter figures, 13.5(b) and 13.5(d) should be compared.

In figure 13.5(a), all the spatial activity can be attributed to thermal fluctuations (stress effects are effectively negligible). This is evidence supporting the nonaffine diffusive processes imagined in FAD. It serves as the baseline for response via single-particle displacements. We expect this to be the primary component of pure (low stress) thermal activation considered in FAD but not STD.

In figure 13.5(b), we see clear presence of bidirectional flow (red-blue interface) as several nucleation events appear along the 45° direction. We regard this to be evidence of free-volume nucleation, a single-particle response threshold attributable primarily to activation by external stress. This is simulation evidence that directly supports the FAD view, and is not excluded by the STD picture. Comparison with figure 13.5(d) shows strong coupling between flow and deformation processes, resulting in the appearance of an emerging shear-band embryo, which is also apparent in figure 13.5(d).

In figure 13.5(c), we see little or no shear transformation activity, which suggests thermal activation plays no role in deformation response as envisaged in STD. A few weak events seem quite random. Comparison with figure 13.5(a) shows the distribution of flow processes at low stress is different from the distribution of deformation processes. We may conclude at low stress, creep proceeds mostly by single-particle rather than cooperative molecular rearrangements.

In figure 13.5(d), we have evidence that at high stress (above a threshold), both diffusive and deformation processes are activated. All the characteristic responses are then observed. There is clearly close correspondence between the spatial distribution of free-volume nucleation (see figure 13.5b) and that of shear-transformation deformation seen here. These are explicit evidence of stress activation.

Summarizing the above for the overall effects of stress on amorphous creep, we apply our observations to predict the variation of creep rate with tensile stress, which is quite widely studied in creep experiments. At low stress, we expect only the effects of thermal activation which, as indicated in figures 13.5(a) and 13.5(c), should lead to a mild increase of creep rate with stress. As stress increases, both flow and deformation mechanisms contribute smoothly until the stress exceeds a threshold at which point thermal and stress activations accelerate. The gap noted in figures 13.3 and 13.4 is an earmark of the crossover behavior expected (see figure 13.6 inset below). Across the gap free-volume nucleation and shear-transformation deformation act cooperatively (positive feedback) and prominently (see figures 13.5b and 13.5d) to cause a sudden increase in creep rate. This is the basis for our anticipation of an upturn in the stress behavior of the creep rate. In the next section, we will directly compare simulation results with experiments so the arguments just given become an explanation based on molecular details that have not been available heretofore.

Creep Rate Upturn

The simulated data given in figure 13.2 allows us to determine a creep rate from the slope of the creep response during steady-state creep (stage 2) at each stress loading. Plotting the results in figure 13.6, we see a stress variation that behaves differently in two regimes. At low-stress, a relatively weak variation is characterized by a creep-rate slope of $n \sim 1.5$, and at high stress a significantly stronger variation with $n \sim 5$, where n is the stress exponent in an expression for the creep rate. The separation in creep-rate behavior is well known in experimental studies; it is widely interpreted as the manifestation

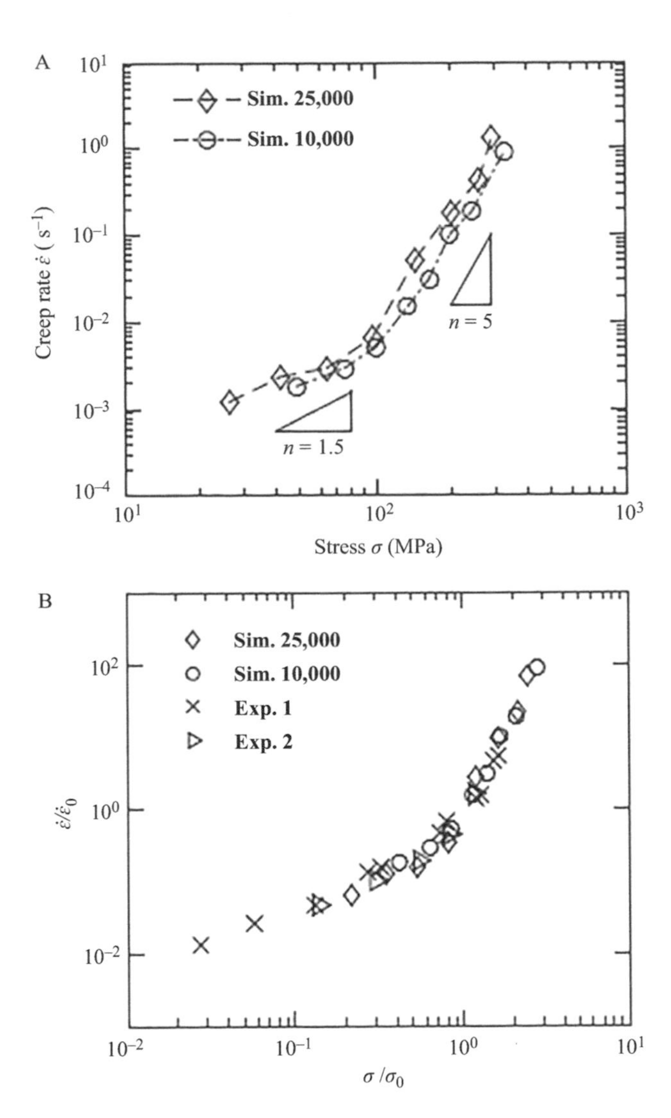

Figure 13.6
Log-log plot of stress variation of creep rate, simulation results at temperature $T = 300$ K (0.78 T_g) derived from the stage-2 part of the creep curve. (a) Simulation results in absolute value at two system sizes (number of atoms). (b) Simulation results and two sets of experimental data, crosses (Lu, Ravichandran, and Johnson 2003) and triangles (Nieh and Wadsworth 2006), all normalized in terms of their threshold values, that is, all curves are scaled to coincide at the point $\varepsilon/\varepsilon_o = 1$, $\sigma/\sigma_o = 1$. Figure from (Cao, Short, and Yip 2017).

of thermal activation processes at low stress changing over to stress activation beyond a characteristic value σ^*.

Notice that the direct comparison of the creep-rate magnitudes would not be possible if simulations were carried out using traditional MD algorithms. The upturn in creep rate indicated by the simulation data matches well between simulation and experiments. Moreover, we observe the characteristic value n^* for the change in index n lies in the vicinity of the stress gap previously noted in the discussions of figures 13.3 and 13.4.

The molecular mechanism associated with the upturn behavior can be traced to a stress activation threshold that we have just identified in discussing figure 13.5, particularly the synergistic response of flow (figure 13.5b) and deformation (figure 13.5d). Thus, we believe it is appropriate to explain the creep-rate response seen in the experiments in terms of the particle flow and rearrangement mechanisms described by the combination of FAD and STD, rather than to support one model over the other.

Mechanism Map

The present study is composed of twenty simulation runs covering the temperatures of 0.5, 0.7, and 0.9 T_g, and stress values ranging from 42 to 258 MPa. In each case, a creep curve has been obtained from which the stress exponent n in stage 2 was determined. The stress-temperature values of our simulations are plotted in figure 13.7, using the symbols open squares and open circles to denote whether the corresponding value of n is ~1 or >5, respectively. Figure 13.7 is therefore in the form of a creep mechanism map. It delineates the stress-temperature regimes where the single-particle diffusion or the shear-transformation deformation predominates.

Looking at the simulation data of squares and circles across the temperature range, one can visualize a mechanism boundary separating the two regimes described by the models proposed by Spaepen (FAD) and Argon (STD), respectively. To motivate such a boundary, we recall two well-known empirical expressions, both based on TST, for the strain rate of an amorphous system in two limiting behavior in the (stress-temperature) phase space. At low stress and high temperature, it is common to write the creep rate in the form

$$\dot{\varepsilon} = \gamma_0 \exp\left(-\frac{Q_d}{kT}\right)\frac{V_d}{KT}\sigma \tag{13.4}$$

where Q_d and V_d are the activation energy and activation volume for the diffusion-dominated process treated in model FAD. In contrast, at high stress the creep rate takes the form

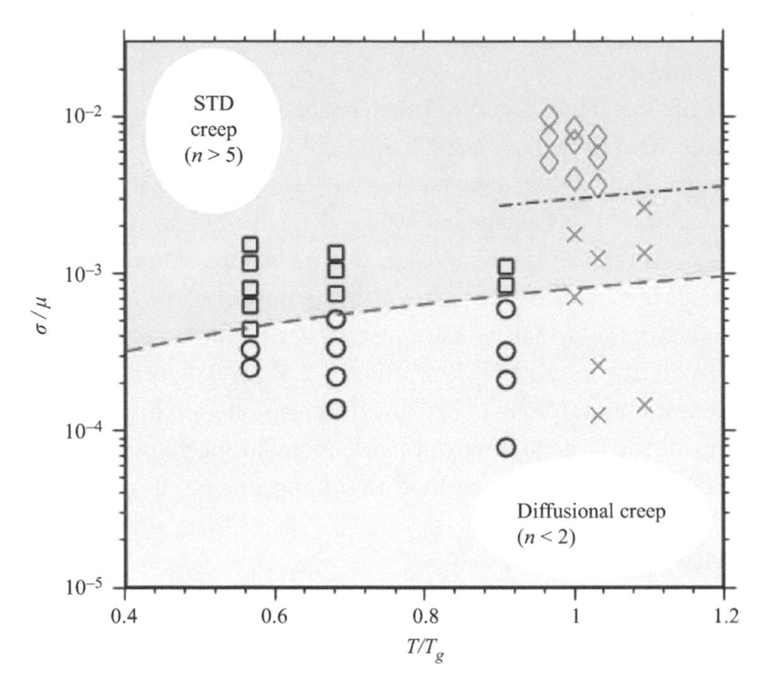

Figure 13.7
Creep mechanism map showing normalized stress-temperature regions of dominance. Simulation data are denoted as open circles ($n<2$) and open squares ($n>5$). Boundary separating predominant shear-transformation deformation mechanism from free-volume atomic diffusion mechanism is indicated by the dashed curve. Also shown are experimental data (Lu, Ravichandran, and Johnson 2003) with open diamond ($n>5$) and crosses ($n<2$). See text for explanation of the red dashed curve. Figure from (Cao, Short, and Yip 2017).

$$\dot{\varepsilon} = \gamma_0 \exp\left(-\frac{Q_{st} - \sigma V_{st}}{kT}\right) = \gamma_0 \exp\left(-\frac{Q_{st}}{kT}\right)\exp\left(\frac{\sigma V_{st}}{kT}\right) \tag{13.5}$$

where the subscript *st* denotes shear transformation.

Setting equation (13.4) equal to equation (13.5), we have an equality involving two sets of activation energy and volume. By treating these quantities parametrically in a numerical analysis, we find that the locus of (σ, T) values is essentially a straight line, $\sigma = a + bT$. The red dashed curve shown in figure 13.7 is therefore drawn having this functional form and adjusted to separate the groups of squares and circles. Thus, the simulation data, without analysis, accommodate a boundary line that has the functional form consistent with equations (13.4) and (13.5). We regard this to be a demonstration of the complementarity of the models of Spaepen and Argon in a unified description of the elementary processes of amorphous plasticity.

Additionally, we include in figure 13.7 mechanical deformation data on a metallic glass (Lu, Ravichandran, and Johnson 2003). Except for an overall shift between the experimental data and simulation results, both point to the same dominant regimes of diffusional versus deformation creep. This is validation that the molecular details we have extracted from our simulations are quite consistent with the system-level responses observed experimentally.

There are several reasons one can give why experiments and simulation should not be expected to match closely in a direct parameter-free comparison. Even when the law of corresponding states is invoked, one should still expect additional effects of mismatch in microstructure imperfections. Since experimental samples necessarily have more defects than the idealized simulation system, this means higher stress loading will manifest in the former, idealized scenario. Another practical reason for expecting no more than only qualitative correspondence is the model metallic glass in simulation is actually a binary mixture and yet its behavior is being compared to measurements on a five-component sample.

Discussion: Dynamical Heterogeneities in Disordered Media

The findings in this essay may be summarized as follows. Two distinct types of elementary processes of amorphous creep are confirmed by probing an experimentally accessible window of waiting time (order of seconds) through molecular simulations. The processes are atomic diffusion through free-volume fluctuations and stress-induced shear transformations. The effects of stress and temperature on the elementary processes are quantified by analyzing and visualizing the spatial fluctuations of the deviatoric strain and nonaffine displacement over an appropriate waiting time interval. At low stress and high temperature, the dominant mechanism is particle flow via thermal fluctuations such as suggested by F. Spaepen. At high stress, the mechanism is a complex interplay of shear transformations, described by A. S. Argon, facilitated by enhanced local free-volume nucleation. Taken together they constitute a mechanistic explanation of the stress and temperature behavior of amorphous creep as experimentally observed.

We have given a unifying picture of the molecular mechanisms of amorphous plasticity, which up to now have been addressed mostly separately through the prevailing models of Spaepen and Argon. The picture is based on molecular details of relevant space-time fluctuations involving both single-particle and collective degrees of the freedom. Our results thus support the notion of a broad class of dynamical order parameters through which one can incorporate existing concepts of dynamical heterogeneities (Kob et al. 1997; Hoffman et al. 2000; Becker and Smith 2003) and variants of self-organized criticality (Jensen 1998).

After clarifying the complementary role of the mechanisms proposed by Spaepen and Argon, an appropriate next step would be to interpret the simulation results in terms of theoretical frameworks of plastic deformation in amorphous materials, such as the STZ concept of a transient flow defect. The original theory (Falk and Langer 1998) has been reviewed (Falk and Langer 2011), and updated (Langer 2015). It appears that STZ can play two different roles: a mechanism combining the ideas of Spaepen and Argon, or a self-contained theoretical framework for interpreting experiments or simulations (Langer 2017; Kamrin and Bouchbinder 2014).

In discussing figure 13.6, we regarded the creep rate upturn as signifying the synergistic actions involving single-particle diffusion and shear-induced deformation. One can imagine quantifying this view by using STZ theory to explicitly calculate the creep-rate response to stress. A similar problem has been studied in which the density of STZs and an effective thermodynamic temperature are introduced as dynamical variables in a set of coupled equations with a stress-dependent deformation rate (Langer 2015). Recalling figure 13.6, we believe the increase of the stress exponent from approximately two to five delineates a regime of nonlinear response, again related to the interplay between diffusion and deformation. More generally, it suggests an interplay between thermal and stress activated processes (Chattoraj, Caroli, and Lemaître 2010).

In summary, our discussions here have revealed the molecular mechanisms of creep in a model metallic glass. At low stress and high temperature, the dominant mechanism is thermally activated particle displacement, while at high stress, it becomes a more complex process of stress-induced enhanced local shear deformation and atomic diffusion. This perspective motivates revisiting the existing notion of dynamical heterogeneities (Kob et al. 1997; Hoffmann et al. 2000; Becker and Smith 2003), as well as a variant of self-organization in a slowly driven threshold system (Jensen 1998), all in the spirit of assessing various related theoretical frameworks, such as the effective thermodynamic temperature (Langer 2015), expanded mode-coupling (Gruber et al. 2016), mean-field with weakening mechanism (Antonaglia et al. 2014), and time-dependent TST modeling (Fan and Cao 2020).

Supplemental Information

We have deferred much of the details of the present simulation study to the original report (Cao, Short, and Yip 2017). For the convenience of the interested reader, we include here a short section on model specification and simulation algorithm.

Consider a two-dimensional $Cu_{50}Zr_{50}$ metallic glass system of size 6.2×31.2 nm. The atoms interact through a Lennard-Jones potential (Krisponeit et al. 2014; Huang et al. 2009) that has previously been used to study plastic deformation (Siebenbürger, Ballauff, and Voigtmann 2012) and thermally activated flows (Schuh, Hufnagel, and Ramamurty 2007). To prepare the initial glassy configuration, we quench a high-temperature liquid at 4×10^{11} K/s at zero pressure using MD with PBCs in all directions. We observe a kink in the volume-temperature curve at 440 K, which will be regarded as the glass transition temperature T_g. After obtaining the amorphous structure, the periodic boundary along the vertical direction is removed to create two free surfaces at top and bottom. Another 400 ps MD simulation is performed to relax the system to zero stress in the NPT ensemble.

To simulate creep using a metadynamics algorithm, we apply a prescribed uniaxial tensile stress and execute the following steps:

1. Perform energy minimization on the equilibrated system to bring it to a local energy minimum.
2. Apply the algorithm ABC (Kushima et al. 2009; Cao et al. 2012) to obtain the TSP and determine the neighboring local minimum state.
3. Compare the internal stress of the new state with the prescribed tensile stress. If the two stresses deviate by more than 1%, perform step 4, otherwise go back to step 2.
4. Perform energy minimization in the presence of the external stress, while allowing the simulation box to vary. Rescale all atom positions whenever the size of simulation box is changed.

The algorithm ABC was designed to circumvent the timescale limitation of traditional MD. It has previously been used in studies of nanocrystalline creep (Lau, Kushima, and Yip 2010), and slow strain rate deformation of defected crystal (Fan et al. 2013) and metallic glasses (Cao et al. 2012). The output of ABC is a set of TSP trajectories, each being an ordered sequence of energy minima and saddle points. The system evolution is then determined by examining the newly sampled configurations of the local energy minima. The activation time of each evolution step is estimated through TST

$$\Delta t_n = \sum_{i=1}^{n}\left[\nu_0 \exp\left(-\frac{\Delta E_i}{k_B T}\right)\right]^{-1}$$

with attempt frequency ν_0 typically taken to be 10^{12} s^{-1}, where ΔE_i is energy barrier of activation path i, and n represents the total number of activation trajectories starting from the initial local minimum.

References

Antonaglia, J., W. J. Wright, X. Gu, R. R. Byer, T. C. Hufnagel, M. LeBlanc, J. T. Uhl, and K. A. Dahmen. 2014. "Bulk metallic glasses deform via slip avalanches." *Physical Review Letters* 112: 155501.

Argon, A. S. 1979. "Plastic deformation in metallic glasses." *Acta Metallurgica.* 27: 47–58.

Becker, T., and J. C. Smith. 2003. "Energy resolution and dynamical heterogeneity effects on elastic incoherent neutron scattering from molecular systems." *Physical Review E* 67 (2): 021904.

Cao, P., M. Li, R. J. Heugle, H. S Park, and X. Lin. 2012. "Self-learning metabasin escape algorithm for supercooled liquids." *Physical Review E* 86: 016710.

Cao, P., M. Short, and S. Yip. 2017. "Understanding the mechanisms of amorphous creep through molecular simulation." *Proceedings of the National Academy of Science* 114 (52): 13631–13636.

Chattoraj, J., C. Caroli, and A. Lemaître. 2010. "Universal additive effect of temperature on the rheology of amorphous solids." *Physical Review Letters* 105: 266001.

Falk, M. L., and J. S. Langer. 1998. "Dynamics of viscoplastic deformation in amorphous solids." *Physical Review E* 57 (6): 7192–7205.

Falk, M. L., and J. S. Langer. 2011. "Deformation and failure of amorphous, solidlike materials." *Annual Review of Condensed Matter Physics* 2 (1): 353–373.

Fan, Y., and P. Cao. 2020. "Long time-scale atomistic modeling and simulation of deformation and flow in solids." In *Handbook of Materials Modeling*, 2nd ed., edited by W. Andreoni and S. Yip. Switzerland AG: Springer Nature.

Fan, Y., Y. N. Osetskiy, S. Yip, and B. Yildiz. 2013. "Mapping strain rate dependence of dislocation-defect interactions by atomistic simulations." *Proceedings of the National Academy of Science* 110 (44): 17756–17761.

Gruber, M., G. C. Abade, A. M. Puertas, and M. Fuchs. 2016. "Active microrheology in a colloidal glass." *Physical Review E* 94: 042602.

Hoffmann, S., L. Willner, D. Richter, A. Arbe, J. Colmenero, and B. Farago. 2000. "Origin of dynamic heterogeneities in miscible polymer blends: a quasielastic neutron scattering study." *Physical Review Letters* 85 (4): 772.

Huang, Y. J., J. Shen, Y. L. Chiu, J. Chen, and J. F. Sun. 2009. "Indentation creep of an Fe-based bulk metallic glass." *Intermetallics* 17 (4): 190–194.

Jensen, H. J. 1998. *Self-Organized Criticality: Emergent Complex Behavior in Physical and Biological Systems*, Vol. 10. Cambridge University Press.

Kamrin, K., and E. Bouchbinder. 2014. "Two-temperature continuum thermomechanics of deforming amorphous solids." *Journal of the Mechanics and Physics of Solids* 73: 269–288.

Kob, W., C. Donati, S. J. Plimpton, P. H. Poole, and S. C. Glotzer. 1997. "Dynamical heterogeneities in a supercooled Lennard-Jones liquid." *Physical Review Letters* 79 (15): 2827.

Krisponeit, J., S. Pitikaris, K. E. Avila, S. Küchemann, A. Krüger, and K. Samwer. 2014. "Crossover from random three-dimensional avalanches to correlated nano shear bands in metallic glasses." *Nature Communications* 5: 3616.

Kushima, A., X. Lin, J. Li, J. Eapen, J. C. Mauro, X. Qian, P. Diep, and S. Yip. 2009. "Computing the viscosity of supercooled liquids." *Journal of Chemistry and Physics* 130: 224504.

Langer, J. 2015. "Shear-transformation-zone theory of yielding in athermal amorphous materials." *Physical Review E* 92: 012318.

Langer, J. 2017. "Yielding transitions and grain-size effects in dislocation theory." *Physical Review E* 95: 033004.

Lau, T. T., A. Kushima, and S. Yip. 2010. "Atomistic simulation of creep in a nanocrystal." *Physical Review Letters* 104 (17): 175501.

Lu, J., G. Ravichandran, and W. L. Johnson. 2003. "Deformation behavior of the Zr41. 2Ti13. 8Cu12. 5Ni10Be22. 5 bulk metallic glass over a wide range of strain-rates and temperatures." *Acta Materialia* 51 (12): 3429–3443.

Nieh, T. G., and J. Wadsworth. 2006. "Homogeneous deformation of bulk metallic glasses." *Scripta Materialia* 54 (3): 387–392.

Schuh, C. A., T. C. Hufnagel, and U. Ramamurty. 2007. "Mechanical behavior of amorphous alloys." *Acta Materialia* 55: 4067–4109.

Siebenbürger, M., M. Ballauff, and T. Voigtmann. 2012. "Creep in colloidal glasses." *Physical Review Letters* 108 (25): 255701.

Spaepen, F. 1977. "A microscopic mechanism for steady state inhomogeneous flow in metallic glasses." *Acta Metallurgica* 25 (4): 407–415.

Weeks, E. R., J. C. Crocker, A. C. Levitt, A. Schofield, and D. A. Weitz. 2000. "Three-dimensional direct imaging of structural relaxation near the colloidal glass transition." *Science* 287 (5453): 627–631.

Further Reading

Cottrell, A. H. 1953. *Dislocations and Plastic Flow in Crystals*. Oxford: Clarendon. Enduring treatise on the subject.

Nabarro, F. R. N., and H. L. Villiers. 1995. *The Physics of Creep*. London: Taylor & Francis. Authoritative discussion on phenomenology and physical mechanisms.

Cottrell, A. H. 1996. "Andrade creep." *Philosophical Magazine Letters* 73: 35. Explanation of 1/3 power law creep as scaling of thermal activations.

Cottrell, A. H. 1996. "Criticality in Andrade creep." *Philosophical Magazine A* 74: 1041. Model to explain transition from critical state to creep regime as localized obstacles giving way, such as avalanche in sandhills reaching a critical slope.

Scherer, G. W. 1996. "Influence of viscoelasticity and permeability on the stress response of silica gel." *Langmuir* 12: 1109. Review of physical factors controlling the response to stimuli such as aging, heating, mechanical deformation, and drying. Ralph Iler award lecture.

Jensen, H. J. 1998. *Self-Organized Criticality*. Cambridge Univ. Press, Cambridge Lecture Notes in Physics. See page 129 for reference to Slowly Driven Interaction Dominated Threshold System (SDIDTS) as fundamental features ascribed to the notion of Self-Organized Criticality of complex systems.

Varnik, F., L. Bocquet, J.-L. Barrat, and L. Berthier. 2003. "Shear localization in a model glass." *Physical Review Letters* 90: 095702. MD simulation showing shear localization in the form of shear bands at low shear rates and stick-slip behavior at very small shear rates.

Shi, Y., and M. L. Falk. 2005. "Strain localization and percolation of stable structure in amorphous solids." *Physical Review Letters* 95: 095502. MD simulation study of the effect of structural relaxation prior to mechanical testing, with more rapidly quenched initial structures undergoing more localization.

Biroli, G. 2007. "A new kind of phase transition?" *Nature Physics* 3: 222. Jamming in 3D binary mixture of glass-forming particles at high densities and slowly driven externally are identified as hallmarks of dynamical heterogeneities.

Chaudhuri, P., L. Berthier, and W. Kob. 2007. "Universal nature of particle displacements close to glass and jamming transitions." *Physical Review Letters* 99: 060604. Distributions of single-particle displacement show the coexistence of slow and fast particles as a signature of dynamical heterogeneities.

Schall, P., D. A. Weitz, and F. Spaepen. 2007. "Structural rearrangements that govern flow in colloidal glasses." *Science* 318: 1895. Visualization of 3D thermally induced particle rearrangements under an applied shear.

Kassner, M. E. 2009. *Fundamentals of Creep in Metals and Alloys*, 2nd ed. Amsterdam: Elsevier. Extensive discussion of mechanisms and data—see pages 70 and 256 on creep rate upturn.

Falk, M. L., and J. S. Langer. 2011. "Deformation and failure of amorphous, solid like materials." *Annual Review of Condensed Matter Physics* 2: 353. Review of development of the STZ model based on nonequilibrium thermodynamics.

Chikkadi, V., and P. Schall. 2012. "Nonaffine measures of particle displacements in sheared colloidal glasses." *Physical Review E* 85: 031402. Confocal microscopy measurements showing power-law spatial correlations at intermediate times implying long-range coupling and critical behavior.

Keralavanna, S. M., T. Cagin, A. Arsenlis, and A. Amine Benzerga. 2012. "Power-Law creep from discrete dislocation dynamics." *Physical Review Letters* 109: 265504. 2D simulation combining dislocation glide and climb to account for matter transport involving vacancy diffusion and coupling to dislocation motion, arriving at creep curves on the timescale of seconds (compare with figures 13.1 and 13.2).

Albe, K., Y. Ritter, and D. Sopu. 2013. "Enhancing the plasticity of metallic glasses: shear band formation, nanocomposites and nanoglasses investigated by molecular dynamics simulations." *Mechanical Materials* 67: 94. Tuning mechanical properties by solute, precipitate, and GB insertions reveals glass-glass interfaces act as structural heterogeneities to promote shear-band formation and prevent strain localization.

Kassner, M. E., K. Smith, and V. Eliasson. 2015. "Creep in amorphous metals." *Journal of Materials Research Technology* 4 (1): 100–107. Review focused on developments since a previous review in 2007.

Sentjabrskaja, T., P. Chaudhuri, M. Hermes, W. C. K. Poon, J. Horbach, S. U. Egelhaaf, and M. Laurati. 2015. "Creep and flow of glasses: strain response linked to the spatial distribution of dynamical heterogeneities." *Science Report* 5: 11884. Using confocal microscopy and MD simulation, a link is established between macroscale creep and microscale single-particle dynamics with regions of enhanced mobilities remaining localized in the creep regime, and strain approximately linearly related to the mean-square displacement.

14 Dynamical Yielding

Discrete stress relaxations (slip avalanches) in a model metallic glass in uniaxial compression are simulated using a metadynamics algorithm at strain rates comparable to experiments. Onset of yielding is directly observed along with spatial and temporal details not yet feasible to obtain with current measurement techniques. Visualization of simulation data reveals that yielding occurs at the first major stress drop, accompanied by the appearance of a single localized shear band region spanning the entire simulation cell. During elastic response prior to yielding, low concentrations of shear transformation deformation events appear spatially uncorrelated and intermittently. During serrated flow following yielding, small stress drops occur interspersed between large drops. The simulation results point to a threshold value of stress dissipation as a characteristic feature separating major and minor avalanches, consistent with mean-field modeling analysis and with mechanical testing experiments. This behavior is interpreted to be the consequence of an interplay of two prevailing mechanisms of amorphous plasticity: thermally activated atomic diffusion and stress-induced shear transformations.

Introduction

Although the plastic response of amorphous solids such as metallic glasses have been of interest for some time, quantitative details of the elementary deformation processes at the nanoscale in space and time are mostly still lacking (Hufnagel, Schuh, and Falk 2016; Cheng and Ma 2011; Schuh, Hufnagel, and Ramamurty 2007). Compression (Antonaglia et al. 2014a; Antonaglia et al. 2014b; Sun et al. 2012) and nanoindentation (Cheng et al. 2014; Schuh and Nieh 2003) experiments have focused on the dynamical evolution of slip avalanches, serrations in the stress-strain behavior of a driven system. A common characteristic of serrated flow is the strain-rate sensitivity becoming more pronounced as rates are lowered. To understand the molecular mechanisms of stress relaxation, as we have seen in the preceding essay, dynamical details available from molecular simulations at appropriate strain rates would be needed. However, as already noted in essays 10 and 12, MD simulations are constrained to strain rates several orders of magnitude higher than the experimental strain rates and typical rates treated in mechanics-based modeling.

Here, we discuss molecular simulation of yielding and serrated flow using the metadynamics algorithm formulated in essay 10 to determine the shear viscosity of supercooled liquids, in essay 12 to probe strain-rate effects, as well as in essay 13 to elucidate the mechanisms of creep. We find that strain rate is an essential fundamental state variable just like temperature, external stress loading, or material microstructure in its effects on the thermomechanical behavior of condensed matter. Thus, essays 12–15 share in common their relevance to materials complexities within the broad concept of dynamical heterogeneities.

Yielding and Serrated Flow

The responses of system-level stress in a 2D metallic glass model subjected to strain rate–controlled uniaxial compression at three characteristic strain rates are shown in figure 14.1 (Cao et al. 2018). The curve QS (quasistatic) is obtained by potential energy

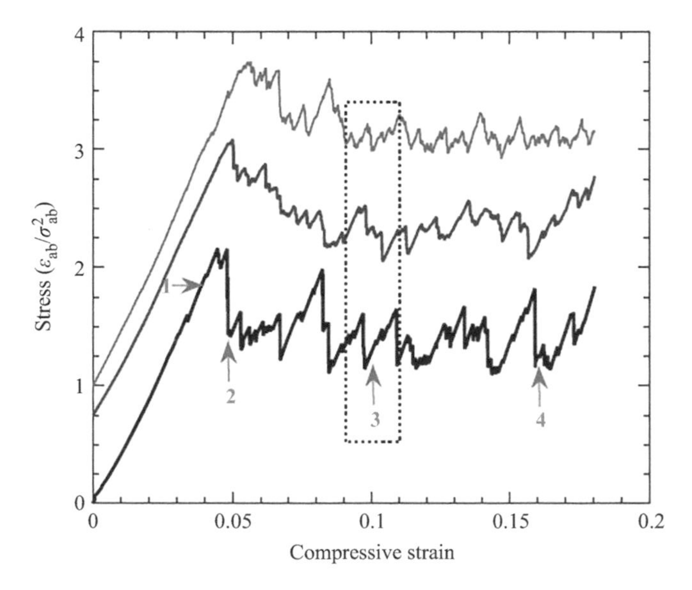

Figure 14.1
Stress-strain response of a model metallic glass under compression at three strain-rate loadings: athermal QS (upper curve), $2.2\times10^{7}\ s^{-1}$ (middle curve), and $4.6\times10^{-2}\ s^{-1}$ (lower curve). The dashed box allows one to appreciate qualitatively the variation in magnitude and frequency of serrated flow over a wide range of strain rates. Arrows indicate the four stages of strain deformation at the lowest strain rate where the atomic configurations are displayed in figure 14.2. Figure from (Cao et al. 2018).

minimization and is therefore the limiting behavior of high strain rate. The other two curves correspond respectively to characteristic strain rate typical of MD simulation, $2.2\times10^7 s^{-1}$, and a strain rate of $4.6\times10^{-2} s^{-1}$, which is comparable to laboratory mechanical tests.

The overall response at each strain rate exhibits initial elastic loading up to the point of yielding, followed by a series of stress relaxations. The effects of strain rate are first a lowering of the entire stress-strain curve as the strain rate decreases from QS to MD to laboratory testing. Secondly, a significant sharpening of the serration occurs in this progression. The variation of the peak (yield) stress, known as stress overshoot, is quite evident. These qualitative features can be intuitively explained as a slower strain rate allows the system more time to respond dynamically through local fluctuations and redistributions at the nanoscale. At the lowest strain rate, we see well-resolved intermittent minor relaxations interspersed between large events throughout the entire flow regime. Similar behavior suggesting a distinction between small and large avalanches is known experimentally (Antonaglia et al. 2014a; Maas, Klaumunzer, and Loffler 2011; Wright, Schwarz, and Nix 2001).

Continuing with figure 14.2, the spatial distributions of local defects are displayed at the lowest strain rate and four system-level strains. These are maps of atomic sites with high deviatoric strain, color coded to indicate the magnitude of D^2_{min}, which is an appropriate measure of local plastic shear transformations (Falk and Langer 1998; Shimizu, Ogata, and Li 2007). The distributions shown are cumulative strains with reference to the initial undeformed configuration at the four stages of evolution indicated by the arrows in figure 14.1. At 4% strain, one sees only a few isolated high-strain sites distributed rather randomly. Immediately after yielding at 4.83%, a band of high-strain sites spanning the system has clearly formed. Also notice the appearance of a surface step. After the appearance of the shear band, subsequent plastic flow is essentially dominated by this localized shear region through thickening and some sliding. For example, during further compression through 10–16%, the band expanded significantly. Correspondingly, the edge steps have become larger to accommodate the additional strain.

Major-Minor Stress Relaxations

To probe more deeply the nanoscale deformation and flow mechanisms associated with an avalanche, two dynamical local order parameters are extracted from the atomic configurations output from the molecular simulations. One is the deviatoric strain D^2_{min} and the other is the nonaffine particle displacement δu. The former involves collective behavior in mechanical response while the latter involves individual particle

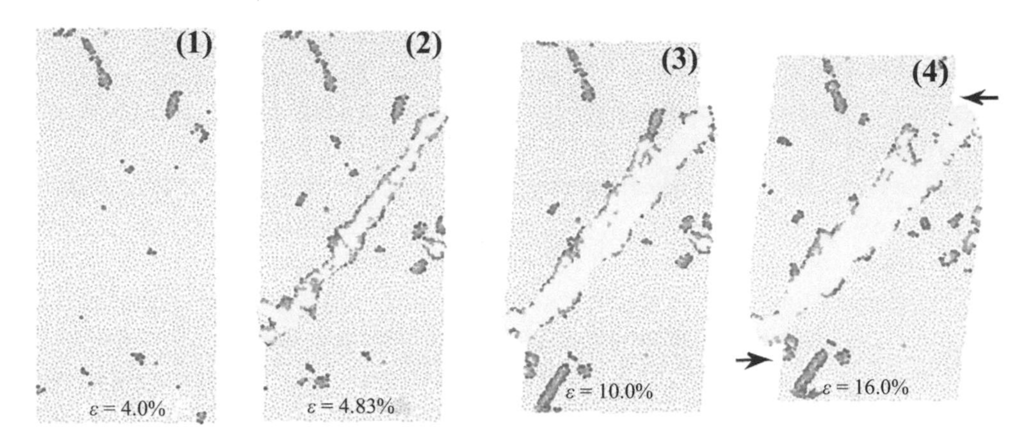

Figure 14.2
Four panels showing local deviatoric strain distributions, color coding ranges from dark red-blue to yellow to a gray background. Overall system strain is (1) 4%, (2) 4.83%, (3) 10%, and (4) 16%. Comparing stages (1) and (2), one sees the formation of an embryonic shear band across the specimen that then thickens with further straining (3) and (4). Also notice surface ledge protrusion during stages (3) and (4). Figure from (Cao et al. 2018).

movements in transport response. Together they represent the total molecular degrees of freedom of the system. Spatial maps of $D^2_{\rm min}$ for small and large stress relaxations are shown in figure 14.3 pertaining to two stages of system deformation at low strain rate. Very little collective effects are involved in a minor avalanche while excitation of $D^2_{\rm min}$ is dominant in a major stress relaxation.

A similar visualization of the nonaffine displacement maps, shown in figure 14.4, reinforces our notion of the mechanistic processes involved in major stress relaxations, namely, single-particle motions also play a significant role together with collective behavior. Given that all the molecular degrees of freedom are activated in a major avalanche, it is not unreasonable to anticipate that their respective fundamental mechanisms are coupled in a nonlinear, mutually reinforcing manner. This would be an example of materials complexities, to be possibly classified as dynamical heterogeneities as we will discuss in the epilogue.

The behavior of deviatoric strain and nonaffine mean-square displacement reaffirms the fundamental distinction between different modeling frameworks of deformation and relaxation in amorphous media, calling for the statistical analysis of the simulation results on the order parameters. Figure 14.5 shows the probability distributions of nonaffine displacement P(δu) separately for the small and large avalanches. Notice that the distribution for minor relaxation is more peaked toward small displacements, the decay following a

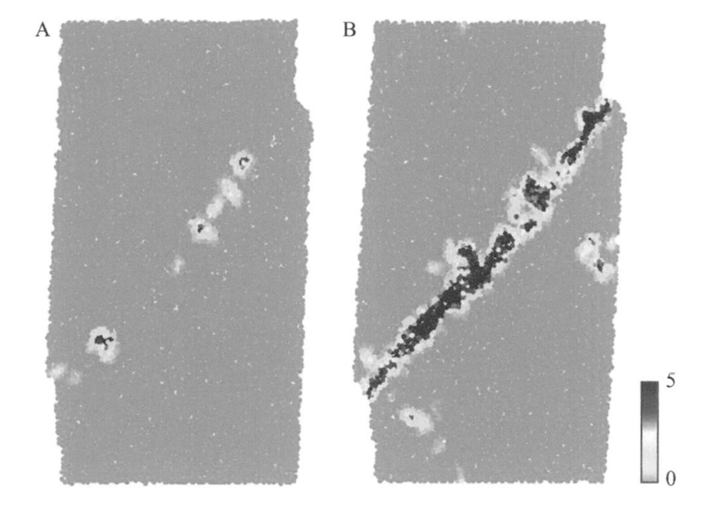

Figure 14.3
Visualization of plastic strain associated with a minor (a) and a major (b) avalanche. Atoms are color coded by the deviatoric strain. The corresponding system strains are 8.3% and 9.7% respectively. The pronounced shear band across the specimen is sustained predominantly by deformations occurring during the major avalanche. Figure from (Cao et al. 2018).

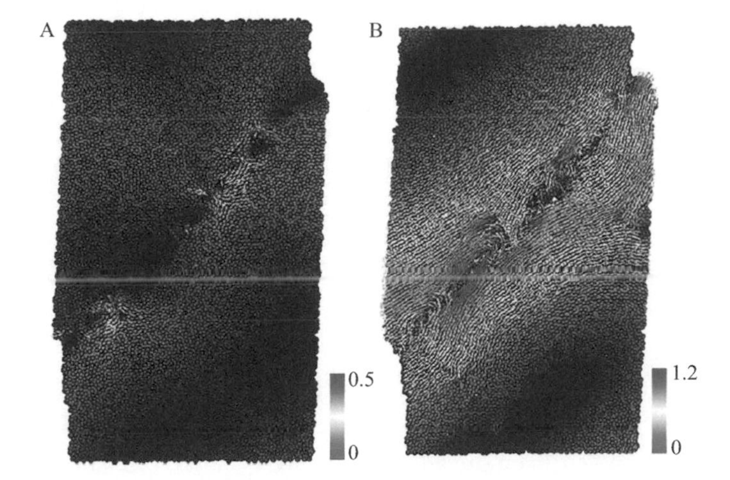

Figure 14.4
Similar to figure 14.3. Visualization of nonaffine displacement associated with a minor (a) and major (b) avalanche. The corresponding system strains are 8.3% and 9.7%, respectively. Notice the scales of color coding are different for the two cases. Figure from (Cao et al. 2018).

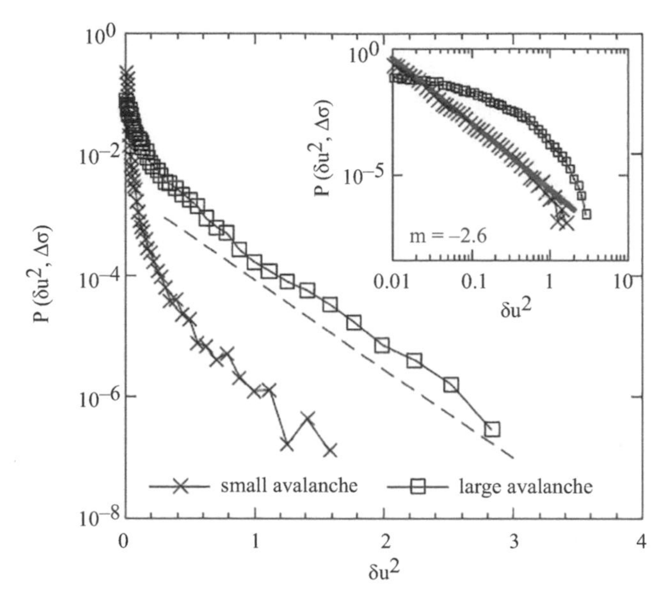

Figure 14.5
Probability distributions of individual particle displacement for small and large avalanches showing different characteristic behavior, power law (exponent –2.6), and exponential tail, respectively. Inset shows log-log scale as opposed to log-linear scale. Figure from (Cao et al. 2018).

power law with an exponent of –2.6 (see figure inset). In contrast, the major relaxations are more broadly distributed with an essentially exponential tail. Similar results have been observed in confocal microscopy experiments on nonaffine fluctuations in sheared colloidal glasses, showing power-law distributions with an exponent of –2.8 (Chikkadi and Schall 2012). MD simulations also have probed the evolution of local deformation in terms of displacement and strain distributions. Exponential tail in displacement distribution has been interpreted to signify contributions from a more strongly sheared region (Maloney and Robbins 2008), essentially what is being discussed here.

In this essay, the metadynamics simulation results point to the existence of a fundamental distinction between small and large serrated flow, as measured by the magnitude of the stress relaxation. Further motivation is provided by cross correlation of the extent of slip (slip size or number of particles participating), the avalanche size $\Delta\sigma$, and the nonaffine mean-square particle displacement δu^2, as given in figure 14.6. Figure 14.6(a) is a scatter plot of the number of atoms estimated to be involved in corresponding avalanches of a given magnitude (size). A break in the slope is clearly evident in the data, meaning larger regions of slip are involved when stress relaxations occur beyond a critical value. This

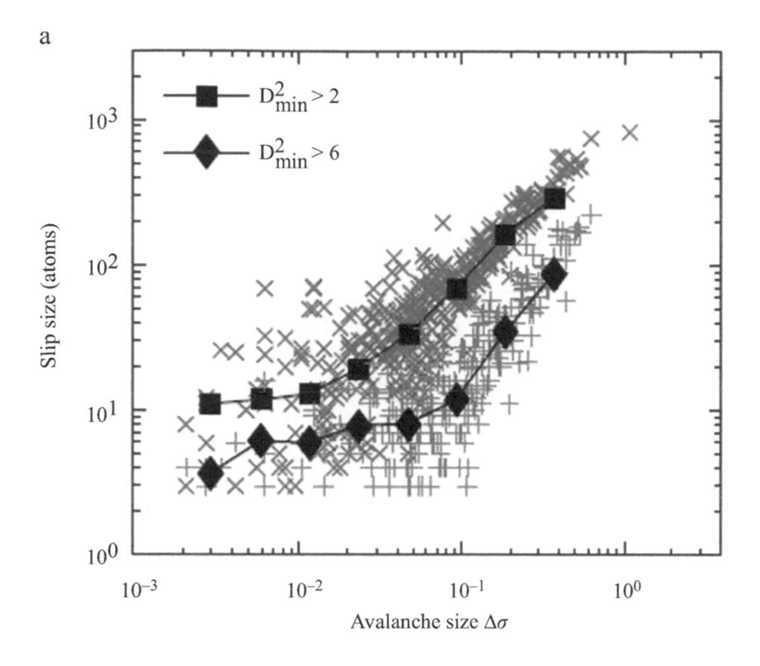

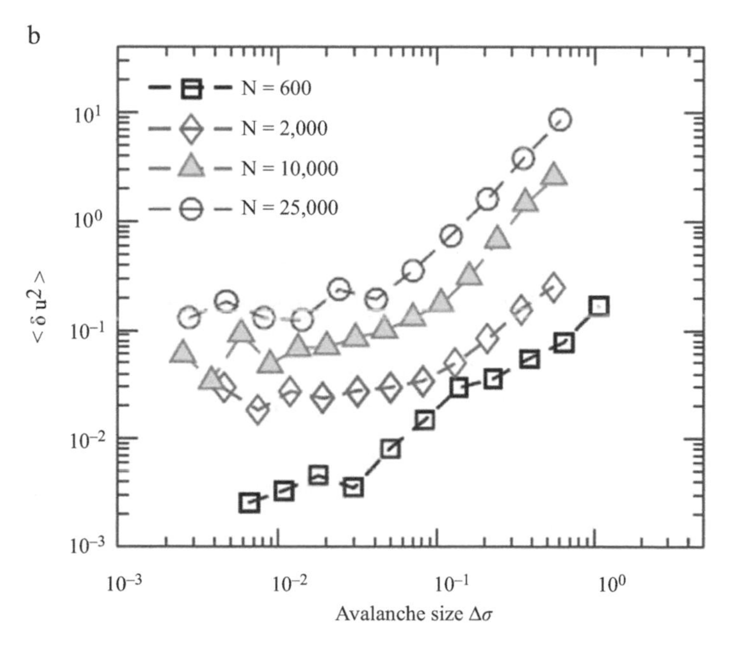

Figure 14.6
Correlation of avalanche size at strain rate of $4.6 \times 10^{-2}\ s^{-1}$ with slip (number of atoms) for system size $N = 10{,}000$ (a), and with nonaffine mean-square displacement at several system size (b). Top three curves in (b) have been shifted vertically to facilitate visualization. Figure from (Cao et al. 2018).

tendency is more pronounced for larger values of deviatoric strain, but is already apparent at the lower value of D^2_{min} threshold. Thus, there is self- consistency in the correlation between greater extent of intense slips and large avalanches. In figure 14.6(b), the results for the two large systems appear to converge at the lower end of $\Delta\sigma$ and also at the higher end. Except for the smallest system of $N = 600$, it appears there is support from the data for the existence of a critical value of avalanche size, and by implication a threshold behavior in the magnitude of stress relaxation in plastic deformation in glassy systems.

The significance of avalanche size has been discussed through a combination of mean-field modeling and high-resolution mechanical testing (Antonaglia et al. 2014a; Wright et al. 2016). Small and large avalanches were interpreted to indicate two distinct modes of plastic deformation, a progressive mode and a more collective mode involving simultaneous shearing, respectively. Both processes are assumed to involve stress relaxations associated with a shear band. Because mechanical testing is much more spatially coarse-grained compared to nanoscale simulation, one should keep this in mind when considering results and interpretations of spatial and temporal correlations across theoretical modeling and analysis, experimental measurements, and nanoscale simulations.

In this essay, we have emphasized the appearance of a shear-band region in the metadynamics simulation of yielding and serrated flow. Shear banding is a ubiquitous mode of deformation in both crystal plasticity and glass rheology; it is particularly of fundamental interest in metallic glasses (Greer, Cheng, and Ma 2013). By quantifying and correlating the atomistic processes during avalanche evolution, we have details to suggest two types of shear-band formation could be at play. One is the percolation of shear transformation events at high strain rate, the other is crack-like propagation at low strain rates. The two modes are demonstrated in figure 14.7. The first mode is observed when small avalanches dominate the mechanical response. Figure 14.7(a) depict the sequence of shear-band formation through the accumulation of many shear transformation events. At the compressive strain of $\varepsilon = 6.85\%$, a number of shear transformation events have been activated with several appearing along the maximum shear stress direction. Upon further step strain of $\Delta\varepsilon = 0.19\%$, the sites begin to merge into an extended region. With another step strain of $\Delta\varepsilon = 0.47\%$, a shear band spanning the entire simulation cell is formed. This scenario of shear band formation in a homogeneous, progressive manner is interpreted to arise from accumulation and percolation of localized plastic deformation events, manifesting at the system level as a series of small avalanches at the strain rates typical of MD simulation.

In contrast to the progressive mode, we find another mode that operates at experimental strain rates. The scenario is an abrupt crack-like extension (recall essay 8). In this case, shear band formation is mediated by large avalanches as shown in figure 14.7(b). At system strain $\varepsilon = 4.80\%$, the localized regions of shear appear in the form of discrete

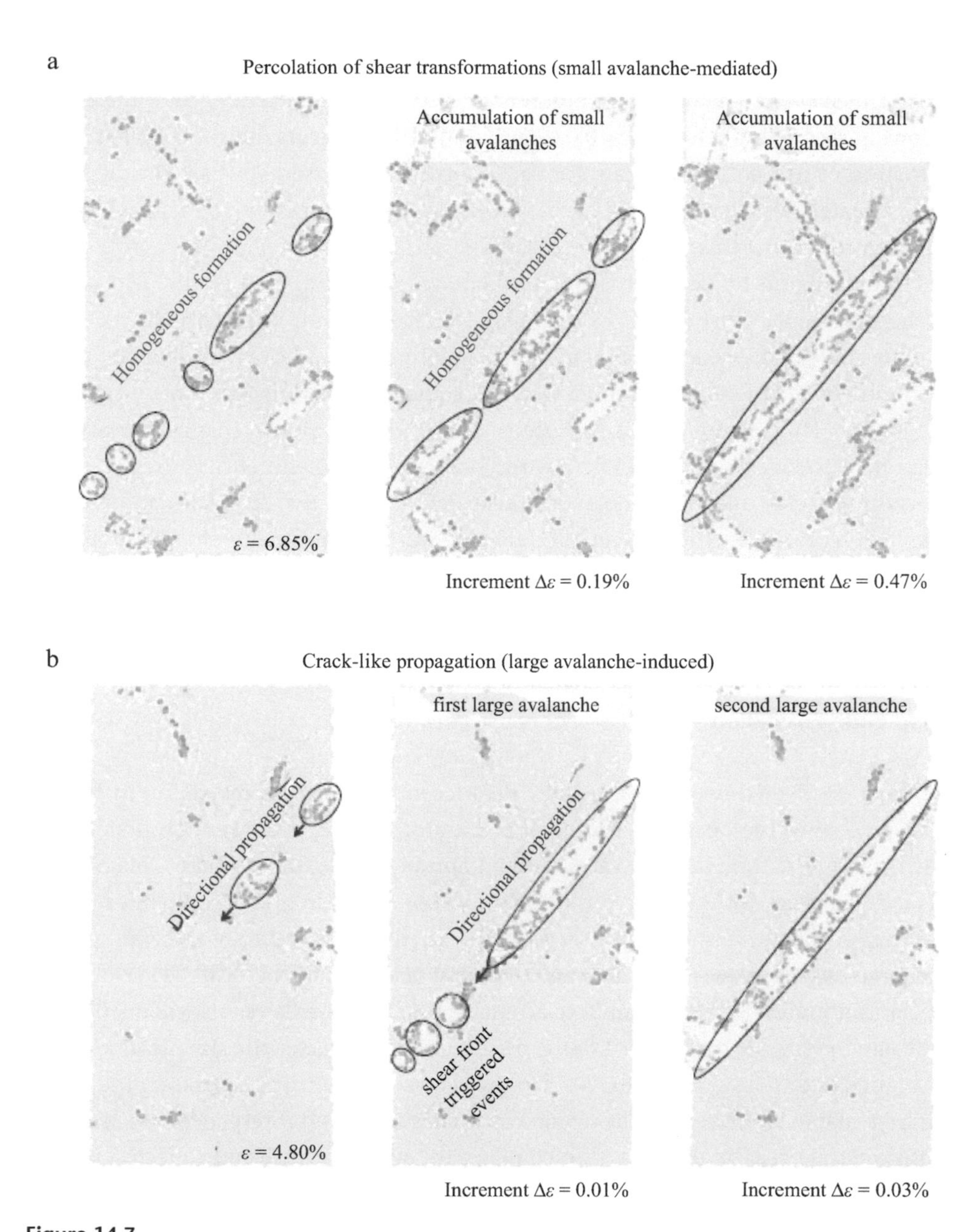

Figure 14.7
Two shear-band formation modes, (a) shear-transformation percolation (b) and crack-like propagation, observed at strain-rate $4.6 \times 10^{-2}\ s^{-1}$. Regions are colored according to local deviatoric strain, and their evolutions are shown with increasing compressive system-level strain and step incremental strains. Figure from (Cao et al. 2018).

shear-weakened sites. At the first large avalanche, which occurs at step increase $\Delta\varepsilon = 0.01\%$, an extended region of shear localization is formed. With the occurrence of the closely following second avalanche ($\Delta\varepsilon = 0.03$ %), the shear band spans the entire system. After formation of this portion of the shear-band, subsequent plastic flow is primarily dominated by large-avalanche activity. We therefore regard such process to be similar to crack-tip extension rather than progressive accumulation.

Through figure 14.7, metadynamic simulations of yielding and plastic deformation provide molecular-level details that support two scenarios of stress relaxations. Recall from figure 14.1 that at low strain rates, only small-scale events are found and no events of intermediate size fall between the largest of the small events and system-spanning size events. Thus, figure 14.1 is consistent with the nucleation scenarios predicted by mean-field theory and high-resolution mechanical test experiments. Other scenarios, concerning preferential nucleation at extrinsic defects and a two-stage formation involving rejuvenation (Cao, Cheng, and Ma 2009), are worthy of future investigations. Besides strain rate, mechanical response can also be tuned with microstructural features such as particle packing fraction in granular materials in dense suspensions (Dahmen, Ben-Zion, and Uhl 2011).

Shear Band Formation

We have uncovered two fundamentally distinct modes of stress relaxation in the serrated flow after the onset of yielding, a preparatory strength-weakening mode where small stress relaxations are activated, and an abrupt release mode where a major stress relaxation occurs without any precursor indications. Yielding is accompanied by the first major avalanche; subsequent serrated flow then proceeds through a series of small and large relaxation events. Spatial and temporal distributions of local deviatoric strain D^2_{min} and nonaffine mean-square displacement $< \delta u^2 >$, as well as statistical distribution of avalanche size $P(\Delta\sigma)$, all point to the existence of a characteristic size $\Delta\sigma_c$ as a feature of serrated flow (Cao et al. 2018). We interpret $\Delta\sigma_c$ to be a measure separating the major-minor avalanches. Moreover, the major avalanches signify an interplay between atomic diffusion and shear localization, thus coupling the single-particle and collective degrees of freedom of the system.

Our recognition of the dynamical coupling between self-diffusion and collective strain distortion implies an opportunity to further interpret existing mechanism models of amorphous plasticity discussed in essay 13. Spaepen proposed atomic diffusion via local free volume to be an order parameter describing the transition from homogeneous to inhomogeneous flow (Spaepen 1977). Argon, on the other hand, proposed the concept of local-stress induced shear transformation as the primary mechanism of plastic deformation in glassy materials (Argon 1979). The nonaffine displacement and the local deviatoric strain distributions analyzed here may be regarded as a way to test the validity of the Spaepen and Argon models. In this respect, our interpretation of the large avalanches as involving coupled diffusion and shear deformation implies a mechanism that combines the two processes. STZ has been introduced to denote a transient flow defect (Falk and Langer 1998). STZ implicitly combines the deformation and diffusion processes of Argon and Spaepen. A more precise relation remains as a topic for future work. Beyond the STZ formulation, the search for a theoretical framework describing physically realistic strain rates is a topic of interest in the community.

References

Antonaglia, J., W. J. Wright, X. Gu, R. R. Byer, T. C. Hufnagel, M. LeBlanc, J. T. Uhl, and K. A. Dahmen. 2014a. "Bulk metallic glasses deform via slip avalanches." *Physical Review Letters* 112: 155501.

Antonaglia, J., X. Xie, G. Schwarz, M. Wraith, J. Qiao, Y. Zhang, P. K. Liaw, J. T. Uhl & K. A. Dahmen. 2014b. Tuned critical avalanche scaling in bulk metallic glasses." *Science Report* 4: 4382.

Argon, A. S. 1979. "Plastic deformation in metallic glasses." *Acta Metallurgica* 27: 47.

Cao, A. J., Y. Cheng, and E. Ma. 2009. "Structural processes that initiate shear localization in metallic glass." *Acta Materialia* 57: 5146.

Cao, P., K. A. Dahmen, A. Kushima, W. J. Wright, H. S. Park, M. P. Short, and S. Yip. 2018. "Nanomechanics of slip avalanches in amorphous plasticity." *Journal of the Mechanics and Physics of Solids* 114: 158.

Cheng, L., Z. Jiao, S. Ma, J. Qiao, and Z. Wang. 2014. "Serrated flow behaviors of a Zr-based bulk metallic glass by nanoindentation." *Journal of Applied Physics* 115: 084907.

Cheng, Y., and E. Ma. 2011. "Intrinsic shear strength of metallic glass." *Acta Materialia* 59: 1800.

Chikkadi, V., and P. Schall. 2012. "Nonaffine measures of particle displacements in sheared colloidal glasses." *Physical Review E* 85: 031402.

Dahmen, K. A., Y. Ben-Zion, and J. T. Uhl. 2011. "A simple analytic theory for the statistics of avalanches in sheared granular materials." *Nature Physics* 7: 554.

Falk, M. L., and J. S. Langer. 1998. "Dynamics of viscoplastic deformation in amorphous solids." *Physical Review E* 57: 7192.

Greer, A., Y. Cheng, and E. Ma. 2013. "Shear bands in metallic glasses." *Materials Science and Engineering. R, Reports* 74: 71.

Hufnagel, T. C., C. Schuh, and M. Falk. 2016. "Deformation of metallic glasses: recent developments in theory, simulations, and experiments." *Acta Materialia* 109: 375.

Maas, R., D. Klaumunzer, and J. Loffler. 2011. "Propagation dynamics of individual shear bands during inhomogeneous flow in a Zr-based bulk metallic glass." *Acta Materialia* 59: 3205.

Maloney, C., and M. O. Robbins. 2008. "Evolution of displacements and strains in sheared amorphous solids." *Journal of Physics* 20: 244128.

Schuh, C. A., and T. Nieh. 2003. "A nanoindentation study of serrated flow in bulk metallic glasses." *Acta Materialia* 51: 87.

Schuh, C. A., T. C. Hufnagel, and U. Ramamurty. 2007. "Mechanical behavior of amorphous alloys." *Acta Materialia* 55: 4067.

Shimizu, F., S. Ogata, and J. Li. 2007. "Theory of shear banding in metallic glasses and molecular dynamics calculations." *Materials Transactions* 48: 2923.

Spaepen, F. 1977. "A microscopic mechanism for steady state inhomogeneous flow in metallic glasses." *Acta Metallurgica* 25: 407.

Sun, B., S. Pauly, J. Tan, M. Stoica, W. Wang, U. Kuhn, and J. Eckert. 2012. "Serrated flow and stick–slip deformation dynamics in the presence of shear-band interactions for a Zr-based metallic glass." *Acta Materialia* 60: 4160.

Wright, W. J., Y. Liu, X. Gu, K. D. Van Ness, S. L. Robare, X. Liu, J. Antonaglia, M. LeBlanc, J. T. Uhl, T. C. Hufnagel, and K. A. Dahmen. 2016. "Experimental evidence for both progressive and simultaneous shear during quasistatic compression of a bulk metallic glass." *Journal of Applied Physics* 119: 084908.

Wright, W. J., R. B. Schwarz, and W. D. Nix. 2001. "Localized heating during serrated plastic flow in bulk metallic glasses." *Materials Science and Engineering. A* 319: 229.

Further Reading

Nemat-Nasser, S., T. Okinaka, and V. Nesterenko. 1998. "Experimental observation and computational simulation of dynamic void collapse in single crystal copper." *Materials Science and Engineering. A* 249: 22. Laboratory-scale materials failure in applied explosive compressive loading showing localization of inelastic flow and subsequent crack propagation.

Cates, M. E., J. P. Wittmer, J.-P. Bouchaud, and P. Claudin. 1998. "Jamming, force chains, and fragile matter." *Physical Review Letters* 81: 1841. Fragility is linked to marginal stability of force-chain networks in the material, unable to support certain incremental loading without plastic rearrangement.

Coussot, P., Q. D. Nguyen, H. T. Huynh, and D. Bonn. 2002. "Avalanche behavior in yield stress fluids." *Physical Review Letters* 88: 175501. Demonstration above a critical stress, gels, clay suspensions and colloidal glasses will flow abruptly and subsequently accelerate leading to avalanches similar to granular materials.

Fuchs, M., and M. Cates. 2002. "Theory of nonlinear rheology and yielding of dense colloidal suspensions." *Physical Review Letters* 89: 248304. A first-principles approach to nonlinear flow of dense suspensions based on an extended mode-coupling theory formalism is shown to capture shear thinning and dynamical yielding.

Rottler, J., and M. O. Robbins. 2003. "Shear yielding of amorphous glassy solids: effects of temperature and strain rate." *Physical Review E* 68: 011507.

Lootens, D., H. Van Damme, and P. Hebraud, 2003. "Giant stress fluctuations at the jamming transition." *Physical Review Letters* 90: 178301. Stress response to a steady imposed shear rate in a concentrated suspension of colloidal particles shows large fluctuations with amplitudes following a power law and well-defined periodicity.

Xue, Q., M. A. Meyers, and V. F. Nesterenko. 2004. "Self-organization of shear bands in stainless steels." *Materials Science and Engineering. A* 384: 35. Study of grain-size effect and shear-band spacing showing only modest variation.

Brader, J. M., T. Voigtmann, M. E. Cates, and M. Fuchs. 2007. "Dense colloidal suspensions under Time-Dependent Shear." *Physical Review Letters* 98: 058301. Extended mode-coupling analysis of driven colloidal suspensions giving a time-dependent description of rheological response far from equilibrium.

Bailey, N. P., J. Schiøtz, A. Lemaître, and K. W. Jacobsen. 2007. "Avalanche size scaling in sheared three-dimensional amorphous solids." *Physical Review Letters* 98: 095501. Athermal QS shear deformation study of spatial scaling.

Dahmen, K. A., Y. Ben-Zion, and J. T. Uhl. 2009. "Micromechanical model for deformation in solids with universal predictions for stress-strain curves and slip avalanches." *Physical Review Letters* 102: 175501. Mean-field theory with one tuning parameter to describe a weakening mechanism.

Brader, J. M., T. Voigtmann, M. Fuchs, R. G. Larson, and M. E. Cates. 2009. "Glass rheology: from mode-coupling theory to a dynamical yield criterion." *Proceedings of the National Academy of Science* 106: 15186. A schematic (single mode) model describing the dynamical yield surface for a class of rheological flows.

Besseling, R., L. Isa, P. Ballesta, G. Petekidis, M. E. Cates, and W. C. K. Poon. 2010. "Shear banding and flow-concentration coupling in colloidal glasses." *Physical Review Letters* 105: 268301. Experiments on hard-sphere colloidal glasses uncover a shear-banding scenario attributed to shear stress–concentration coupling.

Chattoraj, J., C. Caroli, and A. Lemaitre. 2010. "Universal additive effect of temperature on the rheology of amorphous solids." *Physical Review Letters* 105: 266001. Detailed analysis of interplay between loading, thermal activation and mechanical noise shows temperature amounts to lowering the strain at which plastic events occur.

Lau, T., A. Kushima, and S. Yip. 2010. "Atomistic simulation of creep in a nanocrystal." *Physical Review Letters* 104: 175501. First atomistic simulation of stress relaxations showing major and minor energy drops on macroscopic time scales.

Martens, K., and L. Bocquet. 2011. "Connecting diffusion and dynamical heterogeneities in actively deformed amorphous systems." *Physical Review Letters* 106: 156001. Scaling relations between tracer diffusion and flow heterogeneities in amorphous materials.

Kim, J-Y, X. Gu, M. Wraith, J. T. Uhl, K. A. Dahmen, J. R. Greer. 2012. "Suppression of catastrophic failure in metallic glass–polyisoprene nanolaminate containing nanopillars." *Advanced Functional Materials*. 22: 1972–1980. Enhanced mechanical properties reported and attributed to a suppression of discrete strain bursts by appropriate nanolaminate design strategy.

Martens, K., L. Bocquet, and J.-L. Barrat. 2012. "Spontaneous formation of shear bands in a mesoscopic model of flowing disordered matter." *Royal Society of Chemistry* 8: 4197–4205.

Jop, P., V. Mansard, P. Chaudhuri, L. Bocquet, and A. Colin. 2012. "Microscale rheology of a soft glassy material close to yielding." *Physical Review Letters* 108: 148301. Confocal microscopy study connecting rheological "fluidity" with strain-rate tensor to extract a correlation length.

Argon, A. S. 2013. "Strain avalanches in plasticity." *Philosophical Magazine* 93: 3795–3808. Single-crystal results demonstrate universal character of plastic relaxation suggested by the concept of self-organized criticality.

Albe, K., Y. Ritter, and D. Sopu. 2013. "Enhancing the plasticity of metallic glasses: shear band formation, nanocomposites and nanoglasses investigated by molecular dynamics simulations." *Mechanical Materials* 67: 94. Tuning mechanical properties by solute, precipitate, and GB insertions reveals that glass-glass interfaces act as structural heterogeneities to promote shear-band formation and prevent strain localization.

Colombo, J., and E. Del Gado. 2014. "Stress localization, stiffening, and yielding in a model colloidal gel." *Journal of Rheology* 58: 1089. Athermal QS simulation showing stiffening after the elastic regime and before yielding.

Homer, E. R. 2014. "Examining the initial stages of shear localization in amorphous metals." *Acta Materialia* 63: 44. Mesoscale (kinetic MC) simulation of yielding and related shear-band nucleation and propagation, based on the STZ model.

Gruber, M., G. C. Abade, A. M. Puertas, and M. Fuchs. 2016. "Active microrheology in a colloidal glass." *Physical Review E* 94: 042602. Implementation of extended mode-coupling theory describing a driven probe particle in a colloidal glass of hard spheres, comparing calculations with simulations.

Cui, Y., G. Po, and N. Ghoniem. 2016. "Controlling strain bursts and avalanches at the nano- to micrometer scale." *Physical Review Letters* 117: 155502. Dislocation dynamics simulations show strain bursts have scale-free avalanche statistics similar to critical phenomena in general.

Cao, P., M. P. Short, and S. Yip. 2017. "Understanding the mechanisms of amorphous creep through molecular simulation." *Proceedings of the National Academy of Science* 114: 13631. Simulation of stress effects on creep rate in a model metallic glass on experimentally relevant time scales.

Cao, P., M. P. Short, and S. Yip. 2019. "Potential energy landscape activations governing plastic flows in glass rheology." *Proceedings of the National Academy of Science* 116: 18790. Simulation of strain-rate modulated shear-flow regimes in amorphous rheology.

15 Shear-Flow Regimes

Glasses are ubiquitous in their natural response to deformation and flow. Fundamental understanding of the molecular-level mechanisms governing their rheological behavior is a central part of the current research frontier. This essay describes using molecular simulation spanning nearly ten orders of magnitude in the strain rate to probe the atomic rearrangements associated with three characteristic regimes of homogeneous and heterogeneous shear flows. Simulation results combined with modeling analysis reveal distinct scaling behavior in the variation of flow stress with strain rate, signifying a nonlinear coupling between thermally activated diffusion and stress-driven collective dynamics. The emergence of flow heterogeneity is closely correlated with extreme-value distribution of local strain bursts that are not readily accommodated by the immediate surroundings, thereby acting as origins of shear localization. Atomistic mechanisms underlying the flow regimes are interpreted by analyzing a distance matrix of nonaffine particle displacements, yielding evidence of various hopping processes on a fractal potential-energy landscape in which shear transformation and liquid-like regions are triggered by the interplay of thermal and stress activations.

Introduction

Glassy materials flow in response to external mechanical loading and thermal relaxation. Compared to plastic flow of crystalline solids, which is governed by topological defects such as dislocations (Cottrell and Bilby 1949), deformation mechanisms of amorphous materials are less well understood (Hufnagel, Schuh, and Falk 2016). It is generally accepted that shear transformation, where a cluster of atoms embedded in an elastic matrix undergoes inelastic rearrangements (Maloney and Lemaitre 2006; Tanguy, Leonforte, and Barrat 2006) is responsible for the loss of structural stability in an athermal QS process (Argon 1979). Through long-range strain fields (Eshelby 1957), a local distortion would trigger other nearby shear transformation events (Sopu et al. 2017) to give rise to avalanche behavior (Bailey et al. 2007), eventually leading to shear banding (Greer, Cheng, and Ma 2013) (see also essay 14). The accumulation of these plastic events is considered central to the phenomenon of shear localization and macroscopic failure of amorphous solids (Shi and Falk 2005).

Besides mechanical stress, thermally activated processes are also key to understanding the rheological behavior in the deformation of glasses. Amorphous materials typically display complex relaxation spectra (Johari and Goldstein 1970; Debenedetti and Stillinger 2001) consisting of distinct peaks associated with different mechanisms (Pan et al. 2008; Harmon et al. 2007). From the potential-energy landscape point of view, the slow α process is identified as a hopping event escaping a metabasin, whereas the β process corresponds to activation across subbasins within an inherent metabasin (Debenedetti and Stillinger 2001; Harmon et al. 2007; Fan, Iwashita, and Egami 2014). The elementary barrier hopping in the β process is argued to have comparable activation energy to that required to initiate transformation events important for physical properties of aging, rejuvenation, diffusion, and mechanical ductility (Schuh, Hufnagel, and Ramamurty 2007).

The interplay of mechanical loading with thermal stress brings about intriguing rheological behavior in many glassy materials, including colloidal glasses (Berthier and Biroli 2011). Here too, strain rate is a key variable that controls the time window for barrier hopping. Discontinuity in particle diffusion has been observed in the shearing of colloidal glasses, which indicates a transition from homogeneous to heterogeneous flow (Chikkadi et al. 2014). Similar results have been observed in a microscale glass (Ramachandramoorthy et al. 2019), as well as bulk metallic glasses close to the glass transition temperature (Lu, Ravichandran, and Johnson 2003; Nieh and Wadsworth 2006). Detailed studies of the coupling effects between mechanical and thermal activations also have been reported in MD simulations (Chattoraj, Caroli, and Lemaitre 2010; Guan, Chen, and Egami 2010; Cheng and Ma 2011). A concern here regarding comparing simulation and experimental findings is that typical strain rates attained in MD (10^7 s^{-1} or higher) are several orders of magnitude higher than those in laboratory measurements. To compensate for this difference, simulation often resorts to a larger mechanical loading than is physically reasonable. Another consequence of the strain-rate mismatch is that the relatively slow thermal activation processes may not be sampled properly.

Strain Rate–Mediated Shear Flows

We consider a metadynamics simulation study of shear deformation in a 2D binary metallic glass model of Cu_{50} Zr_{50} at temperature $T = 0.75\ T_g$ in the strain-rate controlled mode. For technical details on the simulation protocol and system setup, the reader is referred to the original publication (Cao, Short, and Yip 2019).

Figure 15.1 shows the stress-strain responses at four imposed strain rates. Notice over a range of ten orders of magnitude in strain rate, one sees in figure 15.1(a) the deformation behavior changing from typically solid-like at high strain rates gradually to typically liquid-like at much lower strain rates. To our knowledge, this result constitutes a molecular-simulation based demonstration of tuning the constitutive behavior of a material substance by merely varying the strain rate. Experimentally, it would be analogous to explicitly observing the transition from solid-to-liquid behavior (as in melting) without changing the sample specimen or the measuring apparatus in order to achieve this range of strain-rate resolution.

The first overall impression from figure 15.1(a) is the lowering of the stress response as strain rate decreases (note the stress curve for 20 s^{-1} has been shifted downward by 200 MPA to facilitate visual inspection), an effect known as stress overshoot. Qualitatively comparing the serrations in the four curves, one sees the characteristic behavior of a rigid solid at the highest strain rate changing over to that of a relaxing viscous liquid at the lowest rate where all the serrated features of plastic flow have smeared into a broad band. This is just what one would expect since at low strain rates, the material has enough time to flow and equilibrate (recall the discussions in essay 1 concerning liquid-state dynamics). Figure 15.1(b) shows a magnification of the shear-flow response in the shear strain range of 18–32% to further illustrate how the serration behavior track the change in strain rates. To extract the stress drop size distribution the curves are smoothed through a median filtering with a bin size $\delta\gamma = 0.2$ % (Pratt 1978). Figure 15.1(c) shows the complementary cumulative distribution of stress drop size $F(\delta\sigma)$. The data suggest the distribution follows a power law at high strain rates, while at lower strain rates deviation from avalanche dynamics (Antonaglia et al. 2014; Krisponeit et al. 2014; Lin et al. 2014) is quite pronounced.

Regime-Specific Deformation Mechanisms

The variation of flow stress with strain rate over the wide range from 2.8 s^{-1} to 5×10^{10} s^{-1}, shown in figure 15.2, is a basic motivation for probing the significant aspects of molecular mechanisms governing shear flow. Several distinctive features can be grouped into three regimes. In regime I, the metadynamics simulation results delineate a power law behavior, $\tau \propto \dot{\gamma}^m$, with exponent m being the strain rate sensitivity. At low strain rates in regime I, m is estimated to be 0.75. As strain rate increases, stress grows sublinearly on entering regime II, with m decreasing to ~0.12. This signifies a weakening strain-rate dependence of the flow stress. In regime III, a flow stress upturn occurs as indicated in both metadynamics and MD simulations (figure 15.2b) as well as in previous studies (Chattoraj, Caroli, and Lemaitre 2010).

In view of the low strain rates in regime I, the flow processes are expected to be relatively smooth, and therefore continuum (macroscale) modeling and analyses would be

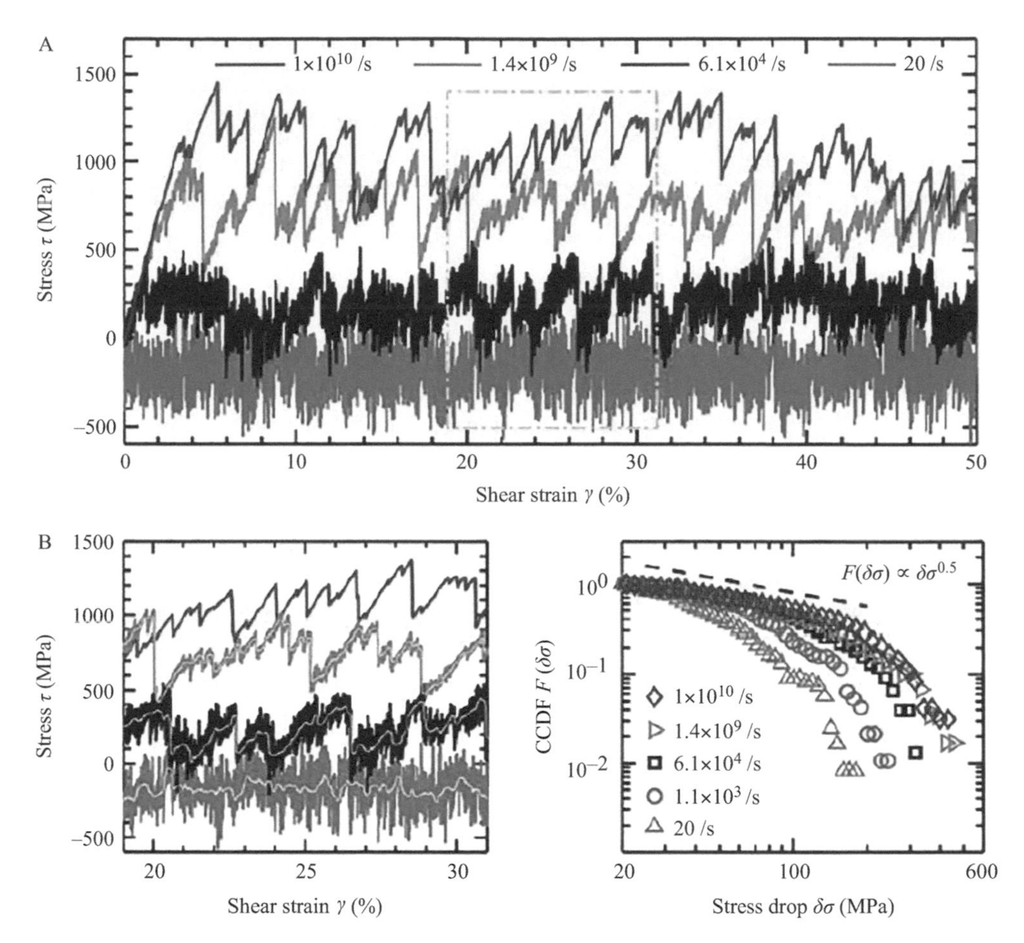

Figure 15.1
Molecular simulation results on the mechanical responses of a 2D metallic glass model in shear at four widely different strain rates (Cao, Short, and Yip 2019). (a) stress-strain curves, color coded according to the strain rates indicated. The curve at the lowest strain rate has been shifted downward by 200 MPa to facilitate visual inspection. (b) Enlargement of (a) in the strain range 18–32%. Yellow lines denote the smoothed curves used to extract data on stress drop events. (c). Complementary cumulative distribution function $F(\delta\sigma)$. Figure from (Cao, Short, and Yip 2019).

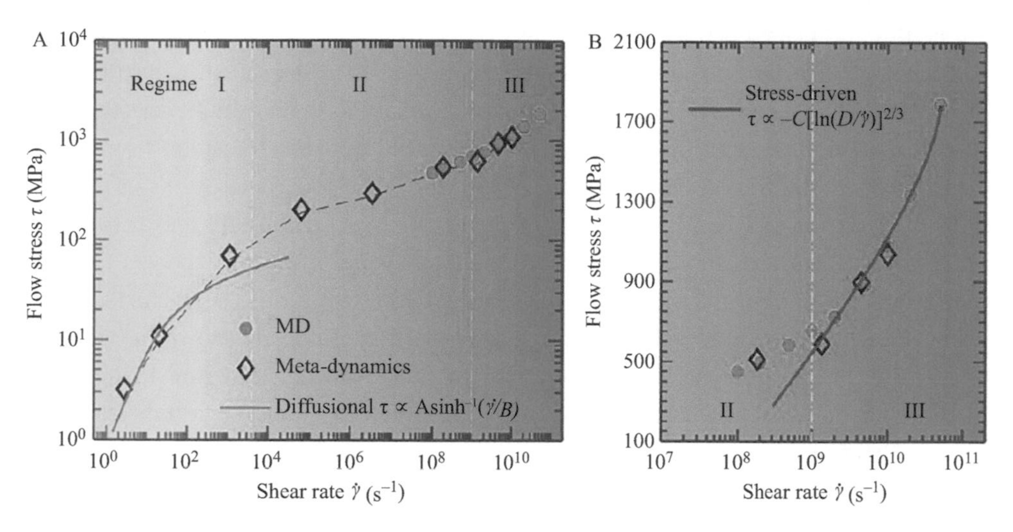

Figure 15.2
Strain rate–modulated regimes of shear-flow stress at temperature $T = 0.75\ T_g$. (a) Log-log plot showing the three regimes spanning ten orders of magnitude in strain rate. Simulation data are shown in symbols. Curve is a continuum modeling prediction. (b) High strain-rate regime III in linear-log plot. Continuum modeling prediction is shown in red. Figure from (Cao, Short, and Yip 2019).

applicable. It is also reasonable to assume a dominant role for thermally activated atomic diffusion. In the presence of low stress, barrier hopping will be biased toward the direction of shear flow. A conventional diffusional plasticity model gives a scaling behavior, $\tau \propto A\sinh^{-1}(\dot{\gamma}/B)$ with A and B being parameters dependent on material and temperature. As seen in figure 15.2(a), this describes well the limiting behavior for the metadynamics simulation results. The agreement is significant in that it provides validation of the simulation capability in a range of strain rates that traditional MD is unable to reach. Beyond regime I, the ABC simulations continue to predict flow stress behavior into regimes II and III. It is significant that in III, figure 15.2(c), there is now agreement between MD and metadynamics. The strain rates here are high enough for MD to be valid, thus this can be regarded as validation of the metadynamics algorithm in addition to experiments. While the convergence of the two types of simulations is to be expected, nonetheless, here the explicit demonstration is gratifying. Also in III, one sees that the upturn behavior is predicted by continuum modeling based on the assumption of stress-induced shear transformation. This means that the upturn behavior known from macroscale modeling (red curve) is now explicitly reproduced and explained by molecular simulations.

Given the flow-regime classification based on metadynamics simulation, we probe further dynamical details from the molecular trajectories. Figure 15.3 is a composite of properties and behavior that one can access across the spectrum of strain rates, namely, spatial maps of dynamical descriptors (order parameters) and their statistical distributions. In figure 15.3(a), we see color-coded maps of local deviatoric strain at three selected strain rates, each belonging in a different regime. One should note the transition from homogeneous distribution of deviatoric strain in regime I to a clearly heterogeneous distribution (appearance of a band of high-strain deformation) in regime III goes through an intermediate regime II, in which pockets of high strain have not fully merged together. This transitory behavior is understandable if the strain rate has not yet reached a critical value. Results of our statistical analysis of local strain magnitude are shown in two ways: probability distribution in figure 15.3(b) and histogram binning in figure 15.3(c). Note the significant variation with strain rate in the distribution of strain magnitude (collective effect). Shear profiles are shown as scatter plots in figure 15.3(d). One sees the expected classical linear profile at low strain rates, and a characteristic kink (bifurcated) profile at high strain rates, the latter having been reported in confocal microscope experiment on a colloidal suspension (Chikkadi et al. 2014). Figure 15.3(e) is the same as figure 15.3(b) with nonaffine particle displacement replacing deviatoric strain. The probability distribution shows a non-Gaussian component that has been ascribed to systems showing dynamical heterogeneity (Chaudhuri, Berthier, and Kob 2007). Non-Gaussian tails becoming more pronounced at higher strain rates means greater participation ratio, which would be consistent with stress-induced shear transformation events.

Nonaffine Distance Matrix

Generally speaking, there is a physical correspondence between varying the strain rate and changing the time interval window of observing an activation event on the potential-energy landscape. To exploit this connection in probing barrier hopping and its corresponding atomic rearrangements, one can attempt to describe the effects of thermal barrier crossing in terms of higher order correlations than just the two-point correlation functions introduced in essay 1. Consider a displacement correlation of the form

$$\Delta^2(\gamma', \gamma'') = \frac{1}{N}\sum_{i}^{N} \left|\mathbf{R}_i(\gamma'') - \mathbf{F}\mathbf{R}_i(\gamma')^2\right| \tag{15.1}$$

where $\mathbf{R}_i(\gamma')$ is the position of atom i at shear strain γ', and the deformation gradient tensor $\mathbf{F}$ is related to the linear affine transformation from γ' to γ''. For a simple shear transformation $\mathbf{F}$ can be expressed as $\mathbf{F} = 1 + (\gamma'' - \gamma')\mathbf{e}_1 \otimes \mathbf{e}_2$, where e_1 and e_2 are basis vectors. We will henceforth call $\Delta^2(\gamma', \gamma'')$ the nonaffine distance matrix (NADM). It is analogous to the distance matrix that has been proposed to identify basin crossing events in thermal

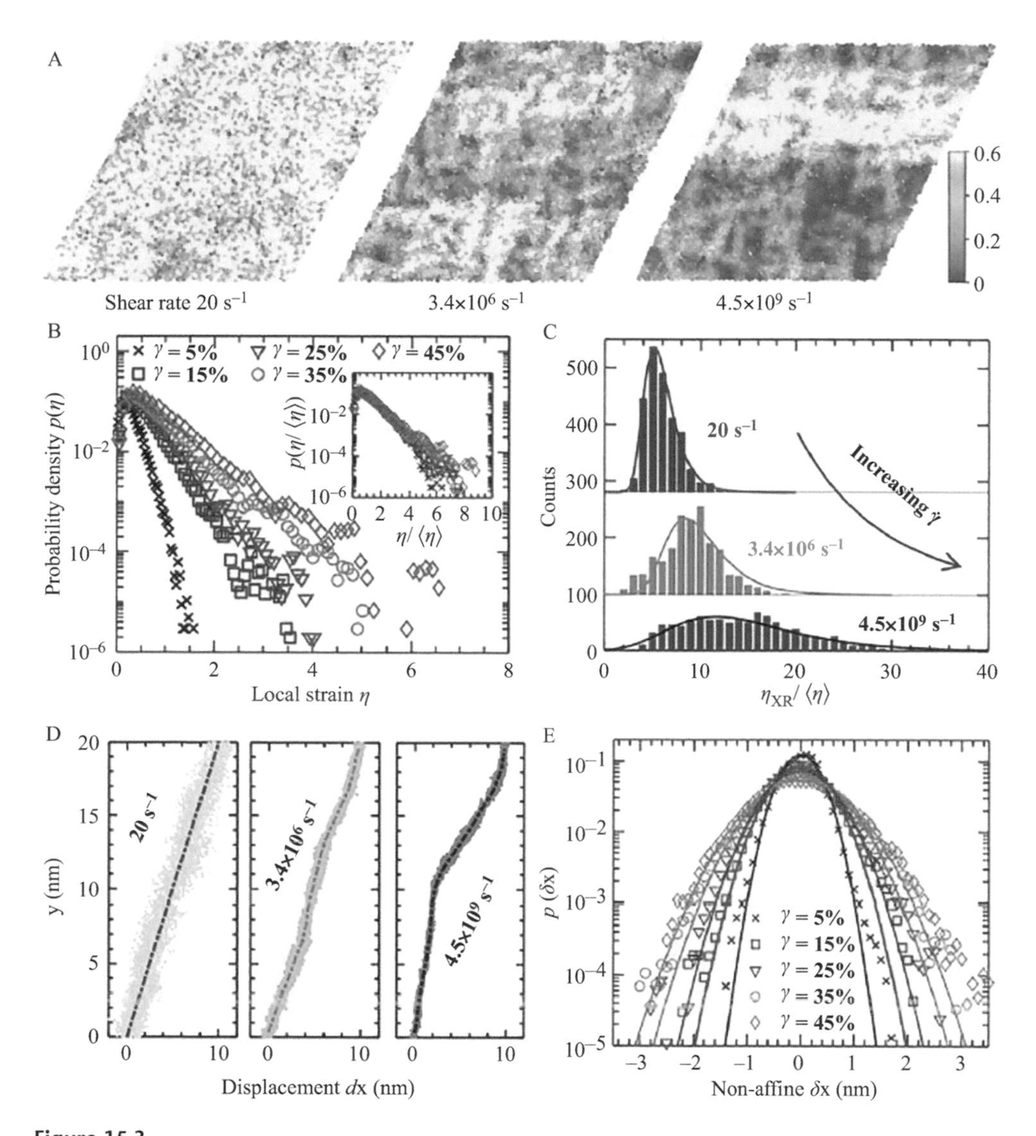

Figure 15.3

Composite of flow fields simulated at three widely separated strain rates. (a) Color-coded spatial maps of deviatoric strain showing homogeneous flow at low strain rates, onset of localized flow at intermediate strain rate, and heterogeneous strain-localized flow at high strain rate. (b) Probability distribution of local deviatoric strain at various values of system-level strain and low strain rate 20 s^{-1}. Inset shows distribution normalized by mean value. (c) Extreme-value distribution analysis of local strain showing the dependence of strain extremes on strain rate. (d) Shear-flow profiles at the three characteristic strain rates. Notice the appearance of a bifurcated (kink) flow at the highest strain rate. Figure from (Cao, Short, and Yip 2019).

activation (Ohmine 1995; Appignanesi et al. 2006). Physically, NADM is a measure of the mean-square nonaffine displacement of a system while its state of strain evolves from γ' to γ''. Although a more complex order parameter than usual, it provides a deeper probe of a system evolution by allowing the single-particle and collective degrees of freedom to be coupled. For example, if the deformation from γ' to γ'' is linearly elastic, the resulting value of NADM would be zero because any local plastic rearrangement due to stress activation or diffusional hopping would leave a finite nonaffine displacement field. In identifying individual barrier hopping and a basin-crossing event, we choose **R** at a local minimum of the PEL, known as the inherent structure (Debenedetti and Stillinger 2001), along the activation trajectory, which should minimize distortions due to thermal vibrations.

Figure 15.4, like figure 15.3, is a composite of NAMD results, which connects the data analysis and interpretation from the macroscale to a schematic of different types of barrier hopping within metabasins. Figure 15.4(a) shows a typical NADM diagram for the strain rate 2.8 s^{-1}. One can see a clustering of darkly shaded (blue) squares, within which the inherent structures are close to each other. There also exist relatively larger squares which are lightly shaded and in which the dark squares are embedded. Note that by magnifying the dark regions, as shown in figure 15.4(b), one finds more shaded blocks appearing and block-to-block crossings, which correspond to transitions among the inherent structures (local PEL minima). Comparing the structures of NADM at different levels reveals a self-similar pattern, which we believe to be significant indications of fractal behavior of NADM. The structure of the NADM graphs suggests that the potential energy basin has a rough bottom consisting of small minima, and the system probably needs to sample several minima before being able to find a pathway to escape.

The multilevel (hierarchical) nature of PEL can be further probed by defining a coarse-grained metric for the distance matrix NADM, $\delta^2(\gamma, \xi) = \Delta^2(\gamma - \xi/2, \gamma + \xi/2)$, which corresponds to a broadening of the diagonal along the strain matrix. Figure 15.4(c) shows the histogram distribution of δ^2 at two extreme strain rates differing by ten orders of magnitude. The value of ξ chosen was smaller than typical strain interval for basin escape but large enough to allow individual hopping between local maxima. Notice the sharp contrast in the two distributions—many high peaks signifying jumps between basins are present at extreme low strain rate compared to only a few discrete low peaks at the extreme high rate. This finding is very suggestive. Between the high peaks in figure 15.4(c) upper curve, the serrated pattern indicates crossings of series of small local minima. On the other hand, in the lower curve, the spectrum is highly discrete. The absence of any baseline activity implies essentially only elastic deformation with no barrier crossing. It seems as if at 10^{10} s^{-1} strain rate, the system is behaving ballistically, impervious to whatever topology is present in the PEL.

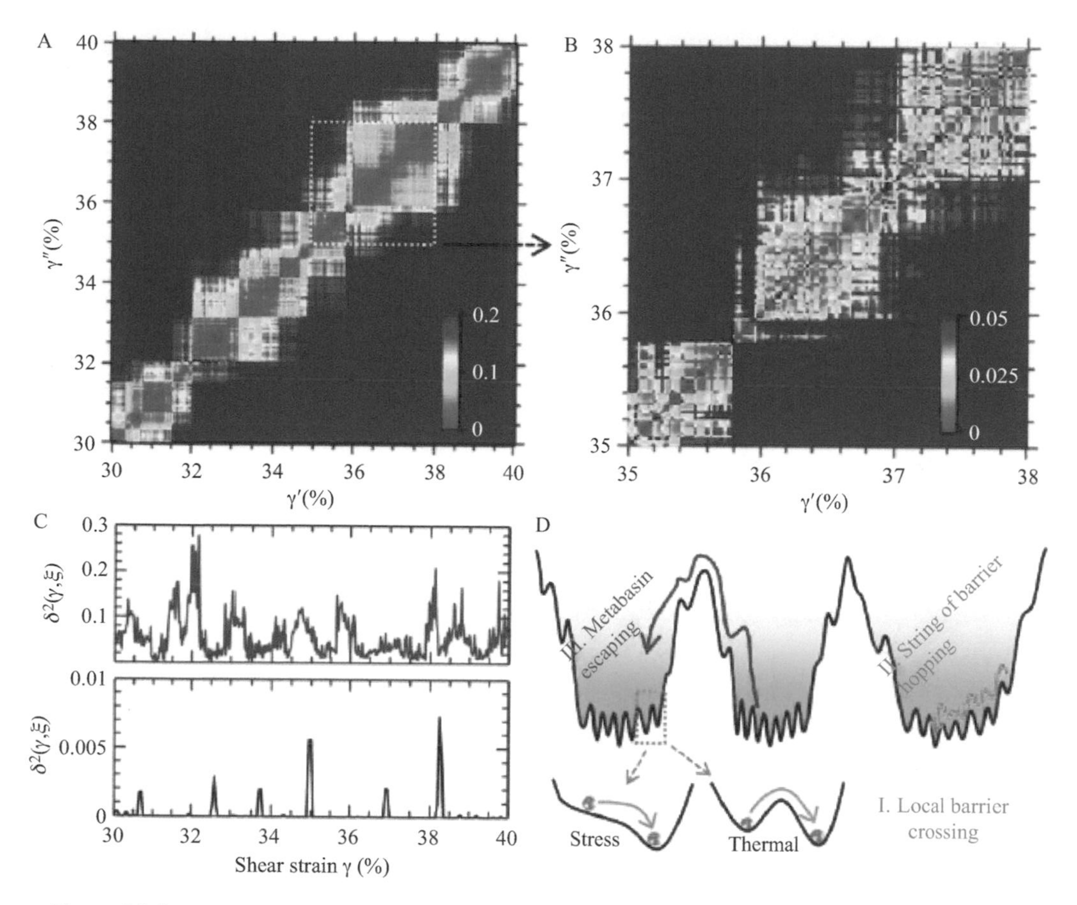

Figure 15.4
Composite of results extracted in NAMD analysis of metadynamics simulation at strain rate of 2.8 s^{-1}. (a) Nonaffine density matrix in the strain range indicated, color coded according the scale shown in the inset. (b) Enlargement of the data inside the dashed white box shown in (a) suggesting self-similarity of the NAMD map. (c) histograms of δ^2 at strain rate 2.8 s^{-1} (upper panel) and 1×10^{10} s^{-1} (lower panel). (d) Schematic of a fractal-like PEL depicting three barrier activation processes within a metabasin (I and II) and between metabasins (III). Figure from (Cao, Short, and Yip 2019).

Physically, it is plausible at strain rate of 2.8 s^{-1} that the system has sufficient time to explore the energy landscape topology and to execute effectively continuous barrier activations. This strain-rate sensitivity is a noteworthy result for the molecular simulation community; it serves as a reminder that timescale issues need to be addressed properly in simulating dynamical phenomena at the molecular level. In the present study, the magnitude of δ^2 was about 0.1, comparable to the value 0.2 for basin transitions in supercooled binary liquids (Appignanesi et al. 2006).

Figure 15.4(d) depicts a schematic of multilevel energy basins to suggest how certain quantitative features of NADM can be interpreted as PEL activations. At constant temperature, the strain rate plays the role of the order parameter governing the mechanism of barrier hopping. For example, it can inform what aspects of the landscape are being visited. Based on these findings, three types of strain rate–mediated activated processes can be identified: crossing of an individual barrier, sequence of small-barrier crossings, and metabasin escape. Activation of a local barrier crossing can result in the loss of local structural stability in the form of a shear transformation event. It retains an athermal nature even at finite temperature if the deformation is stress dominated, particularly at elevated strain rates.

Glass Rheology Complexity

To characterize the spatial nature of the mechanisms underlying the three characteristic types of barrier hopping, one can consider again maps of the local deviatoric strain to see how the local defects are clustered in different strain-rate regimes. This is shown in figure 15.5.

Figure 15.5(a) is a scenario where only a few isolated local shear transformation events corresponding to local barrier crossing have been activated. This is typically the situation of early deformation (low system-level strain). In figure 15.5(b), as defect formation

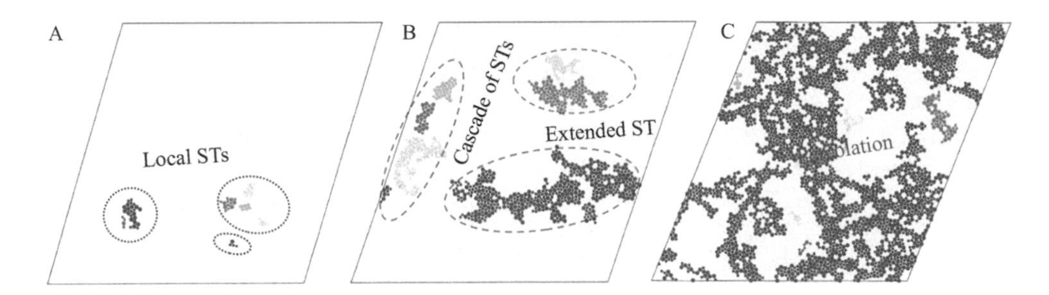

Figure 15.5
Defect distribution maps sampled in metadynamics simulation of shear flow that illustrate three characteristic types of barrier activation processes. This figure can be rationalized from the three regimes of shear flow mediated by strain rates, figure 15.2. (a) Local barrier activation involving localized shear transformations. (b) A string of correlated barrier hopping leading to an extended shear transformation region and a cascade event, with chain-shape clustering. Color coding is by cluster size with dark red representing the largest. (c) Metabasin escape resulting in percolation of liquid-like segments toward a system spanning cluster. The seven largest clusters are shown whose constituent atoms have plastic strains greater than 10%. Figure from (Cao, Short, and Yip 2019).

increases with increasing strain rate in regime II, neighboring shear transformations are triggered biased primarily along the direction of shear—note color coding in figure 15.5(b). Upon entering regime III, figure 15.5(c), the rapid appearance of plasticity events amounts to a kind of jamming (percolative) process resulting in system-level heterogeneous shear flow—compare with the rightmost panel in figure 15.3(a). The series of concerted local barrier hopping within a broad PEL basin (metabasin) could occur at intermediate strain rates when the system has more time to explore more local minima. For example, in figure 15.5(b), it can be seen that various processes are activated, including chain-like and extended shear transformations resulting from local elastic cage breakup. The third type of activation is related to the escape from one metabasin to an adjacent one, a process central to the dynamics of the glass transition (Debenedetti and Stillinger 2001) (see also essay 11). When a metabasin hopping occurs—crossing over a large peak in δ^2—a high concentration of liquid-like sites is nucleated and these percolate throughout the systems as seen in figure 15.5(c). For metabasin activation, it is found the atoms have activation probability 0.62, which is above the bond percolation threshold 0.5 (Stauffer and Aharony 1992). Thus, a cluster spanning the system could occur. The percolation of these activated soft spots is therefore likely to be the mechanism responsible for the homogeneous flow in regime I. Recalling the earlier discussion of transition from solid-to-liquid–like behavior in figure 15.1(a), the mechanism illustrated in figure 15.5(c) provides an explanation of the behavior of shear-induced fluidization in slowly deformed colloidal glass (Eisenmann et al. 2010).

Theoretical models considering various molecular mechanisms of glass deformation and flow have been proposed to address the issue of inhomogeneous plasticity. Existing models include free-volume assisted atomic diffusion (Spaepen 1977), stress-induced shear transformation (Argon 1979), and STZ (Falk and Langer 1998). The relevance of each model in terms of its ability to quantitatively describe the three flow regimes can now be examined, along with how its implementation can be integrated into the scenario of thermal and stress activation from a PEL perspective. The results discussed here suggest the predominant process in regime I is percolation of liquid-like soft defects via thermally activated diffusion, and primarily shear transformation deformation in regime III. In regime II thermal and stress activations significantly influence each other, and need to be treated as coupled complex (nonlinear) mechanisms. While the present study is focused on strain rate–mediated rheological behavior at a temperature below the glass transition, it would be quite worthwhile to ask for more details of crossover mechanisms across regimes. It is anticipated at lower temperature that the crossover strain rates would shift to lower values simply because the hopping time for a barrier crossing would increase exponentially with decreasing temperature (according to TST). Above the glass transition temperature, it seems reasonable to speculate that the distinction between Regimes I and II could vanish as the system now lies outside of a fractal deep metabasin (Hwang, Riggleman, and Crocker 2016).

In the PEL scenario, the local processes of deformation and diffusion appear as saddle-point activations in the evolving system. While the detailed physical nature of the barriers involved depends on the specific phenomenon under discussion, it is clear that the accommodating mechanisms governing the different responses will require a framework of multilevel energy basins and associated nesting of saddle points inherent in a topologically complex surface. For example, the landscape scenario plays a significant role in understanding the fundamental nature of the glass transition as illustrated by the temperature variation of the shear viscosity of supercooled liquids (Debenedetti and Stillinger 2001; Goldstein 1969; Angell 1995) (see also essay 11). Deep in the glass state, it is believed there exists a roughness transition in connection with phase transition in spin glasses (Gardner 1985). This has been discussed theoretically for structural glasses using a disordered ensemble of hard spheres leading to the interpretation of a metabasin breaking up into a hierarchy of subbasins (Charbonneau et al. 2014). The fractal feature of the energy landscape has been predicted to exist in other glassy materials (Hwang, Riggleman, and Crocker 2016). The present perspective on the PEL is based on the features of NADM and δ^2. By performing a scaling analysis, one obtains estimates of the fractal dimension of 1.69 and 1.36, respectively (Cao, Short, and Yip 2019). Since these are different measures of the underlying PEL, one may conclude that both the metabasin area and its barrier height can exhibit fractal behavior, which would also apply to figures 15.4 and 15.5 in their relevance regarding atomic rearrangements, manifestations of avalanches, and rheological responses.

As a final note, rheological behavior mediated by strain rates is a phenomenon that illustrates a level of materials complexity currently being referred to as dynamical heterogeneities. Such a scenario depicts metallic glasses as an assembly of soft spots or liquid-like regions (free-volume excitations) (Manning and Liu 2011; Ding et al. 2014; Ding et al. 2016). This point of view is very much aligned with figure 15.5 regarding spatial distributions and with figures 15.2 and 15.3 regarding strain-rate resolution. Concerning the three regimes of shear flow, we can expect at high strain rates, regime III, there will be a wide distribution of local strain and triggering of extreme values. At intermediate strain rates, regime II, the coupling of thermal activation and the associated atomic diffusion gives rise to string-like soft-spot activation or elongated cascades. If the strain rate is further lowered to regime I, the coupled thermal-stress activations leads to formation of liquid-like regions via collective rearrangements, and if these percolate through the system, flow in a homogeneous manner would ensue. Taking the three regimes together, one may interpret the transitional behavior of regime II to be the synergistic consequence of thermal and stress activations. In terms of atomic rearrangement processes, it is also a recognition of the interplay between stress-induced shear transformation and thermally activated atomic diffusion. An additional unifying aspect of the present perspective is a suggestion to emphasize the equivalence among the current theoretical frameworks describing amorphous plasticity. Different candidates seem to be based on different principles, such as thermodynamics-based STZ theory (Falk and Langer 2011; Langer 2008), time correlation functions in the self-consistent mode coupling approximation (Gruber et al. 2016), and variants of mean-field models in condensed-matter theory (Dahmen, Ben-Zion, and Uhl 2009).

References

Angell, C. A. 1995. "Formation of glasses from liquids and biopolymers." *Science* 267: 1924–1935.

Antonaglia, J., W. J. Wright, X. Gu, R. R. Byer, T. C. Hufnagel, M. LeBlanc, J. T. Uhl, and K. A. Dahmen. 2014. "Bulk metallic glasses deform via slip avalanches." *Physical Review Letters* 112: 155501.

Appignanesi, G. A., J. A. Rodriguez Fris, R. A. Montani, and W. Kob. 2006. "Democratic particle motion for metabasin transitions in simple glass formers." *Physical Review Letters* 96: 057801.

Argon, A. S. 1979. "Plastic deformation in metallic glasses." *Acta Metallurgica* 27: 47–58.

Bailey, N. P., J. Schiøtz, A. Lemaitre, and K. W. Jacobsen. 2007. "Avalanche size scaling in sheared three-dimensional amorphous solids." *Physical Review Letters* 98: 095501.

Berthier, L., and G. Biroli. 2011. "Theoretical perspective on the glass transition and amorphous materials." *Reviews of Modern Physics* 83: 587–645.

Cao, P., M. P. Short, and S. Yip. 2019. "Potential energy landscape activations governing plastic flows in glass rheology." *Proceedings of the National Academy of Science* 116: 18790.

Charbonneau, P., J. Kurchan, G. Parisi, P. Urbani, and F. Zamponi. 2014. "Fractal free energy landscapes in structural glasses." *Nature Communications* 5: 3725.

Chattoraj, J., C. Caroli, and A. Lemaitre. 2010. "Universal additive effect of temperature on the rheology of amorphous solids." *Physical Review Letters* 105: 266001.

Chaudhuri, P., L. Berthier, and W. Kob. 2007. "Universal nature of particle displacements close to glass and jamming transitions." *Physical Review Letters* 99: 060604.

Cheng, Y. Q., and E. Ma. 2011. "Intrinsic shear strength of metallic glass." *Acta Materialia* 59: 1800–1807.

Chikkadi, V., D. M. Miedema, M. T. Dang, B. Nienhuis, and P. Schall. 2014. "Shear banding of colloidal glasses: observation of a dynamic first-order transition." *Physical Review Letters* 113: 208301.

Cottrell, A. H., and B. A. Bilby. 1949. "Dislocation theory of yielding and strain ageing of iron." *Proceedings of the Physics Society of London A* 62: 49–62.

Dahmen, K. A., Y. Ben-Zion, and J. T. Uhl. 2009. "Micromechanical model for deformation in solids with universal predictions for stress-strain curves and slip avalanches." *Physical Review Letters* 102: 175501.

Debenedetti, P. G., and F. H. Stillinger. 2001. "Supercooled liquids and the glass transition." *Nature* 410: 259–267.

Ding, J., Y.-Q. Cheng, H. Sheng, M. Asta, R. O. Ritchie, and E. Ma. 2016. "Universal structural parameter to quantitatively predict metallic glass properties." *Nature Communications* 7: 13733.

Ding, J., S. Patinet, M. L. Falk, Y. Cheng, and E. Ma. 2014. "Soft spots and their structural signature in a metallic gas." *Proceedings of the National Academy of Science* 111: 14052–14056.

Eisenmann, C., C. Kim, J. Mattsson, and D. A. Weitz. 2010. "Shear melting of a colloidal glass." *Physical Review Letters* 104: 035502.

Eshelby, J. D. 1957. "The determination of the elastic field of an ellipsoidal inclusion, and related problems." *Proceedings of the Royal Society of London A* 241: 376–396.

Falk M. L., and J. S. Langer. 1998. "Dynamics of viscoplastic deformation in amorphous solids." *Physical Review E* 57 (6): 7192–7205.

Falk, M. L., and J. S. Langer. 2011. "Deformation and failure of amorphous, solid like materials." *Annual Review of Condensed Matter Physics* 2: 353–373.

Fan, Y., T. Iwashita, and T. Egami. 2014. "How thermally activated deformation starts in metallic glass." *Nature Communication* 5: 5083.

Gardner, E. 1985. "Spin glasses with *p*-spin interactions." *Nuclear Physics B* 257: 747–765.

Goldstein, M. 1969. "Viscous liquids and the glass transition: a potential energy barrier picture." *Journal of Physics and Chemistry* 51: 3728–3739.

Greer, A. L., Y. Q. Cheng, and E. Ma. 2013. "Shear bands in metallic glasses." *Materials Science and Engineering. R: Report* 74: 71–132.

Gruber, M., G. C. Abade, A. M. Puertas, and M. Fuchs. 2016. "Active microrheology in a colloidal glass." *Physical Review E* 94: 042602.

Guan, P., M. Chen, and T. Egami. 2010. "Stress-temperature scaling for steady-state flow in metallic glasses." *Physical Review Letters* 104: 205701.

Harmon, J. S., M. D. Demetriou, W. L. Johnson, and K. Samwer. 2007. "Anelastic to plastic transition in metallic glass-forming liquids." *Physical Review Letters* 99: 135502.

Hufnagel, T. C., C. A. Schuh, and M. L. Falk. 2016. "Deformation of metallic glasses: recent developments in theory, simulations, and experiments." *Acta Materialia* 109: 375–393.

Hwang, H. J., R. A. Riggleman, and J. C. Crocker. 2016. "Understanding soft glassy materials using an energy landscape approach." *Nature Materials* 15: 1031.

Johari, G. P., and M. Goldstein. 1970. "Viscous liquids and the glass transition. II. Secondary relaxations in glasses of rigid molecules." *Journal of Chemistry and Physics* 53: 2372–2388.

Krisponeit, J.-O., S. Pitikaris, K.E. Avila, S. Küchemann, A. Krüger, and K. Samwer. 2014. "Crossover from random three-dimensional avalanches to correlated nano shear bands in metallic glasses." *Nature Communications* 5: 3616.

Langer, J. R. 2008. "Shear-transformation-zone theory of plastic deformation near the glass transition." *Physical Review E* 77: 021502.

Lin, J., E. Lerner, A. Rosso, and M. Wyart. 2014. "Scaling description of the yielding transition in soft amorphous solids at zero temperature." *Proceedings of the National Academy of Science* 111: 14382–14387.

Lu, J., G. Ravichandran, and W. L. Johnson. 2003. "Deformation behavior of the Zr41. 2Ti13. 8Cu12. 5Ni10Be22. 5 bulk metallic glass over a wide range of strain-rates and temperatures." *Acta Materialia* 51: 3429–3443.

Maloney, C. E., and A. Lemaitre. 2006. "Amorphous systems in athermal, quasistatic shear." *Physical Review E* 74: 016118.

Manning, M. L., and A. J. Liu. 2011. "Vibrational modes identify soft spots in a sheared disordered packing." *Physical Review Letters* 107: 108302.

Nieh, T. G., and J. Wadsworth. 2006. "Homogeneous deformation of bulk metallic glasses." *Scripta Materialia* 54: 387–392.

Ohmine, I. 1995. "Liquid water dynamics: collective motions, fluctuation, and relaxation." *Journal of Physics and Chemistry* 99: 6767–6776.

Pan, D., A. Inoue, T. Sakurai, and M. W. Chen. 2008. "Experimental characterization of shear transformation zones for plastic flow of bulk metallic glasses." *Proceedings of the National Academy of Science* 105: 14769–14772.

Pratt, W. K. 1978. *Digital Image Processing*. New York: John Wiley & Sons.

Ramachandramoorthy, R., J. Schwiedrzik, L. Petho, C. Guerra-Nuñez, D. Frey, J.-M. Breguet, and J. Michler. 2019. "Dynamic plasticity and failure of microscale glass: rate-dependent ductile–brittle–ductile transition." *Nano Letters* 19: 2350–2359.

Schuh, C. A., T. C. Hufnagel, and U. Ramamurty. 2007. "Mechanical behavior of amorphous alloys." *Acta Materialia* 55: 4067–4109.

Shi, Y., and M. L. Falk. 2005. "Strain localization and percolation of stable structure in amorphous solids." *Physical Review Letters* 95: 095502.

Sopu, D., A. Stukowski, M. Stoica, and S. Scudino. 2017. "Atomic-level processes of shear band nucleation in metallic glasses." *Physical Review Letters* 119: 195503.

Spaepen, F. 1977. "A microscopic mechanism for steady state inhomogeneous flow in metallic glasses." *Acta Metallurgica* 25: 407–415.

Stauffer, D., and A. Aharony. *Introduction to Percolation Theory,* 2nd ed. Taylor & Francis.

Tanguy, A., F. Leonforte, and J.-L. Barrat. 2006. "Plastic response of a 2D Lennard-Jones amorphous solid: detailed analysis of the local rearrangements at very slow strain rate." *European Physics Journal E* 20: 355–364.

Further Reading

Laughlin, R. B., D. Pines, J. Schmalian, B. P. Stojković, and P. Wolynes. 2000. "The middle way." *Proceedings of the National Academy of Science* 97: 32. Mesoscopic organizations in various forms of matter point to limits of current understanding of fundamental organizing principles and the emergence of new research frontiers.

Rodney, D., and C. Schuh. 2009. "Distribution of thermally activated plastic events in a flowing glass." *Physical Review Letters* 102: 235503.

Wagner, N., and J. F. Brady. 2009. "Shear thickening in colloidal dispersions." *Physics Today* 27. Stiffening of complex fluids under shear-stress impact.

Lematre, A., and C. Caroli. 2009. "Rate-dependent avalanche size in athermal sheared amorphous solids." *Physical Review Letters* 103: 065501. $T = 0$ finite strain-rate study of spatial scaling.

Kushima, A., J. Eapen, J. Li, S. Yip, and T. Zhu. 2011. "Time scale bridging in atomistic simulation of slow dynamics: viscous relaxation and defect activation." *European Physics Journal B* 82: 271. Viscous relaxation and defect activation in glassy states.

Chikkadi, V., and P. Schall. 2012. "Nonaffine measures of particle displacements in sheared colloidal glasses." *Physical Review E* 85: 031402. Confocal microscopy measurements showing power-law spatial correlations at intermediate times implying long-range coupling and critical behavior.

Yip, S., and M. P. Short. 2013. "Multiscale materials modelling at the mesoscale." *Nature Materials* 12: 774. Slow-dynamics phenomena at the mesoscale science research frontier.

Wyart, M., and M. E. Cates. 2014. "Discontinuous shear thickening without inertia in dense non-Brownian suspensions." *Physical Review Letters* 112: 098302. Two distinct types of DST: above the jamming point of frictional particles and a second regime at lower densities associated with strain-rate hysteresis.

Yip, S. 2016. "Understanding the viscosity of supercooled liquids and the glass transition through molecular simulations." *Molecular Simulation* 42: 1330. Understanding the viscous behavior of supercooled liquids through molecular simulations.

Cao, P., M. P. Short, and S. Yip. 2017. "Understanding the mechanisms of amorphous creep through molecular simulation." *Proceedings of the National Academy of Science* 114: 13631. Molecular mechanisms of creep on experimental time scales unveiled and unified.

Cao, P., K. A. Dahmen, A. Kushima, W. J. Wright, H. S. Park, M. P. Short, and S. Yip. 2018. "Nanomechanics of slip avalanches in amorphous plasticity." *Journal of the Mechanics and Physics of Solids* 114: 158.

Scalliet, C., B. Giuslin, and L. Berthier. 2022. "Thirty milliseconds in the life of a supercooled liquid." *Physical Review X* 12: 041028. Relaxation dynamics of model supercooled liquids simulated by molecular dynamics over a time range of 10 orders of magnitude.

Epilogue: Toward Materials Complexity: Dynamical Heterogeneities

Essay Highlights

After the final essay, we come to a concluding section where the salient points discussed throughout this book are very briefly summarized to give an outlook. In correspondence with the essay overviews in the prologue, we show a table of significance and broad insights in the same format. Interested readers may wish to regard table P.1 in the prologue and table E.1 below as complementary overviews on the essays' significance for readers before and after reading the book.

New Age of Data Science and Societal Relevance

Rising interests in artificial intelligence (AI) and machine learning (ML) are driving fundamental shifts in practically all aspects of science and technology globally. As in changes that directly impact societal welfare, such shifts present unprecedented challenges and opportunities. In October 2018, the Massachusetts Institute of Technology (MIT) received a major gift to establish a new College of Computing. MIT decided to leverage this exceptional opportunity to foster Artificial Intelligence (AI)-breakthroughs across all the disciplines pursued at the Institute. With a budget of $1 billion, a new building, and fifty additional faculty positions, MIT believed it could play a leadership role in AI teaching and research, not only in multidisciplinary science and applications but also in their integration with ethics.

The MIT AI initiative aimed to accomplish two goals. One is to solve the problem of overloading (excessive student demand) the Computer Science part of the Department of Electrical Engineering and Computer Science (EECS). The other is to enhance the integration of ethics in computing and data science and technology across the Institute. While MIT is highly ranked for its Computer Science program, it may not be as widely known that the Institute is also highly ranked in the social sciences among

Table E.1
Mechanisms and physical insights addressed by each essay.

1	Space-time correlations, particle collisions, hydrodynamics, kinetic and memory effects
2	MD simulation, unique features of atomistic simulation
3	Mode-coupling theory, nonlinear feedback, particle localization in diffusion
4	Crystal-melting elasticity criteria, MD test, phase transitions triggered in series
5	Melting in Si, defect-surface effects, thermodynamics versus kinetics
6	Stress-induced plasticity, dislocation nucleation and mobility, twinning and kink mechanisms
7	Shear deformation in Cu and Al, charge-density effects, shearability of metals and ceramics
8	Reaction pathway sampling, crack tip nucleation and extension, brittle-ductile behavior
9	Nanocrystal strength, intragranular to intergranular crossover, reverse Hall–Petch scaling
10	Transition-state metadynamics simulation, supercooled liquid viscosity
11	Molecular nature of glass transition, fragile-strong viscosity, energy-landscape activation
12	Transition-state modeling, strain-rate effects in crystal plasticity, glass-rheology analogy
13	Amorphous creep, ubiquitous upturn, stress localization, mechanism map
14	Dynamical yielding, major-minor relaxations, shear-band formation scenarios
15	Strain rate–modulated shear-flow regimes, metabasin energy-landscape activations

universities worldwide. Even without a school of medicine or law, there are many aspects of ethics that MIT can address through the implications and responsibilities of computing.

Looking forward at the science and technology community broadly, one might ask how the rise of AI/ML would impact the computational materials community. This issue has already received considerable attention in several contributions in the second edition of the *Handbook of Materials Modeling* (Andreoni and Yip 2020). One view is that computational materials is entering into a new age of Data Science (Marzari 2020; Draxl and Scheffler 2020). That this trend is certain to grow in scope and participation is quite clear from the program emphasis in almost all the upcoming conferences and workshops. An observation here is there clearly exists an opportunity to bridge the cultural and domain expertise gap between the traditional computer science and the computational science communities, the latter being well represented by computational materials (see figure P.2 for an analogy with the MSS frontier). Since data can naturally include simulation results, it means that there will be a significant role for materials mechanisms to play in enhancing the reliability of materials AI through the paradigm of computational thinking (Wing 2008; Ananthaswamy 2021). Another thought on the evolution of computational modeling and simulation is to introduce event consequences as a metric of success in addressing social-societal issues such as climate change, energy sustainability, and global conflict resolution (Kusnezov 2007).

An Outlook

Future prospects of computational materials, in terms of science and technology opportunities, are indeed very bright. There will be no shortage of global challenges in the arenas of climate change, energy sustainability, infrastructure renewal, nuclear arms control, and so on. Materials mechanisms invariably will be at the core of these activities and discussions. We leave the reader with a specific problem, admittedly only one among many well-qualified candidates that combine scientific bandwidth with societal urgency.

The problem is cement setting. Shown in figure E.1 is the time evolution of the strength of cement paste. When cement powder is mixed with water at a construction site, the paste is poured into a mold (a building column or bridge support) to be

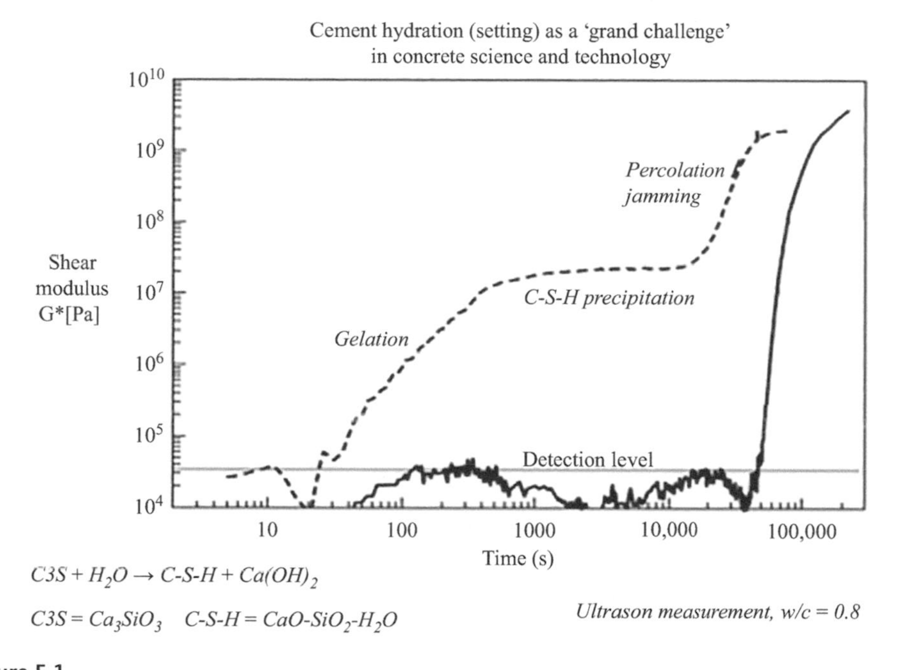

Figure E.1
Cement setting curves (experimental) (Lootens et al. 2004). Time development of shear modulus showing three stages: gelation (first increase, dashed curve), incubation (plateau), and setting (second increase). With the addition of polymer plasticizers, the gelation and incubation stages essentially disappear, leaving only the setting stage (solid curve). The three-stage behavior may be compared with the creep curve figure 13.1 in essay 13. Figure adapted from (Lootens et al. 2004).

hardened overnight. The mechanism of cement hydration or setting, especially the third stage during which the strength increases from 10 MPa to 1 GPa, is an unresolved challenge with direct relevance to climate change (CO_2 emission), energy sustainability, and infrastructure renewal. It is also a materials-centric issue for multi-investigator collaborations (Concrete Sustainability Hub 2021; Pellenq et al. 2009; Masoero et al. 2012; Van Vliet et al. 2012; Yip and Short 2013; Pinson et al. 2015; Ioannidou et al. 2016).

From the perspective of this book, a close connection naturally exists between materials mechanisms and the emerging role of AI/ML. Indeed, one can draw an analogy with the MSS frontier that lies between the micro- and the macroscales (see figure P.2 in the prologue). In delineating MSS, we have previously pointed to three problems of slow dynamics to illustrate the fundamental role of mechanisms: the temperature variation of the viscosity of supercooled liquids, the mechanical setting of cement paste, and the phenomenon of stress corrosion cracking (Yip and Short 2013). While each one is a distinct phenomenon of longstanding interest, together they span a domain of complex materials behavior ripe for further exploration. Add to this group the present discussions of creep, yielding, and shear flow (the essays grouped in part V), we have then identified a research arena one may call *Dynamical Heterogeneities* (Berthier et al. 2011). This seems to be surely a fertile research frontier for the computational materials community.

References

Ananthaswamy, A. 2021. "Jeannette Wing." *MIT Technology Review* 124(6): 64.

Andreoni, W., and S. Yip, eds. 2020. *Handbook of Materials Modeling,* 2nd ed. Switzerland: Springer Nature.

Berthier, L., G. Biroli, J.-P. Bounchaud, L. Cipelletti, and W. van Saarloos. 2011. *Dynamical Heterogeneities in Glasses, Colloids, and Granular Media,* Vol. 150. Oxford: Oxford University Press.

Concrete Sustainability Hub. 2021. *A Scientific Investigation into Concrete Pavement Durability.* Report of the MIT Concrete Sustainability Hub.

Draxl, C., and M. Scheffler. 2020. "Big Data-Driven Materials Science and Its FAIR Data Infrastructure." In *Handbook of Materials Modeling,* 2nd ed., edited by W. Andreoni, and S. Yip. Switzerland: Springer Nature.

Ioannidou, K., K. J. Krakowiak, M. Bauchy, C. G. Hoover, E. Masoero, S. Yip, F.-J. Ulm, P. Levitz, R. J.-M. Pellenq, and E. Del Gado. 2016. "Mesoscale texture of cement hydrates." *Proceedings of National Academy of Science* 112: 2029.

Kusnezov, D. 2007. "How big can you think?" *Computing in Science Engineering* IEEE and A. I. P., 88.

Lootens, D., P. Hebrud, E. Lecolier, and H. Van Damme. 2004. "Gelation, shear-thinning and shear-thickening in cement slurries." *Oil and Gas Science and Technology* 59: 31.

Marzari, N. 2020. "Materials Informatics: overview." In *Handbook of Materials Modeling,* 2nd ed., edited by W. Andreoni, S. Yip. Switzerland: Springer Nature.

Masoero, E., E. Del Gado, R. J. M. Pellenq, F.-J. Ulm, and S. Yip. 2012. "Nanostructure and nanomechanics of cement: polydisperse colloidal packing." *Physical Review Letters* 109: 155503.

Pellenq, R. J.-M., A. Kushima, R. Shashavri, K. J. Van Vliet, M. J. Buehler, S. Yip, and F.-J. Ulm. 2009. "A realistic molecular model of cement hydrates." *Proceedings of the National Academy of Science* 106: 16102.

Pinson, M. B., E. Masoero, P. A. Bonnaud, H. Manzano, Q. Ji, S. Yip, J. J. Thomas, M. Z. Bazant, K. J. Van Vliet, and H. M. Jennings. 2015. "Hysteresis from multiscale porosity: modeling water sorption and shrinkage in cement paste." *Physical Review Applied* 3: 064009.

Van Vliet, K. J., R. J. M. Pellenq, M. J. Buehler, J. C. Grossman, H. Jennings, F.-J. Ulm and S. Yip. 2012. "Set in stone? A perspective on the concrete sustainability challenge." *MRS Bulletin* 37: 395.

Wing, J. 2008. "Computational thinking and thinking about computing." *Philosophical Transactions of the Royal Society A* 366: 3717.

Yip, S., and M. P. Short. 2013. "Multiscale materials modelling at the mesoscale." *Nature Materials* 12: 774.

Further Reading

Bak, P., C. Tang, and K. Wiesenfeld. 1987. "Self-Organized Criticality: an explanation of $1/f$ noise." *Physical Review Letters* 59: 381. Attempt to explain the source of $1/f$ as the dynamics of a critical state as a universal phenomenon.

Bak, P., C. Tang, and K. Wiesenfeld. 1988. "Self-Organized Criticality." *Physical Review A* 38: 464. Certain extended dissipative dynamical systems naturally evolve into a critical state, with no characteristic time or length scales, temporal and spatial signatures being $1/f$ noise and scale-invariant fractal structure, respectively.

Bak, P. 1996. *How Nature Works.* New York: Springer-Verlag. Development of the notion of Self-Organized Criticality (SOC) in statistical physics.

Jensen, H. J. 1998. *Self-Organized Criticality.* Cambridge Univ. Press, Cambridge Lecture Notes in Physics. See page 129 for reference to Slowly Driven Interaction Dominated Threshold System (SDIDTS) as fundamental features ascribed to the notion of Self-Organized Criticality of complex systems.

Chaudhuri, P., L. Berthier, and W. Kob. 2007. "Universal nature of particle displacements close to glass and jamming transitions." *Physical Review Letters* 99: 060604. Distributions of single-particle displacement show the coexistence of slow and fast particles as a signature of dynamical heterogeneities.

Biroli, G. 2007. "A new kind of phase transition?" *Nature Physics* 3: 222. Jamming in 3D binary mixture of glass-forming particles at high densities and slowly driven externally are identified as hallmarks of dynamical heterogeneities.

Zhang Y., M. Lagi, F. Ridi, E. Fratini, P. Baglioni, E. Mamontov, and S. H. Chen. 2008. "Observation of dynamic crossover and dynamic heterogeneity in hydration water confined in aged cement paste." *Journal of Physics: Condensed Matter* 20: 502101. Quasielastic neutron scattering study of the temperature variation of translational relaxation time indicating the presence of a crossover temperature with dynamical heterogeneity implications.

Filoche, M., and M. Mayboroda. 2012. "Universal mechanism for Anderson and weak localization." *Proceedings of the National Academy of Science* 109: 14761. Localization of stationary waves in mechanical, acoustical, optical and quantum systems associated with an inhomogeneous medium, complex geometry, or quenched disorder manifested by partitions into weakly coupled subregions.

Ding, J., S. Patinet, M. L. Falk, Y. Cheng, and E. Ma. 2014. "Soft spots and their structural signature in a metallic gas." *Proceedings of the National Academy of Sciences* 111: 14052. Demonstrates the correlation between soft vibrational modes associated with the most disordered local polyhedral packing environments and STZ composed of atoms with large nonaffine displacements, effectively a statistical coupling of topological heterogeneity with vibrational-relaxational heterogeneity.

So, K. P., M. Stapelberg, Y. R. Zhou, M. Li, M. P. Short, and S. Yip. 2022. "Observation of dynamical transformation plasticity in metallic nanocomposites through a precompiled machine-learning algorithm." *Materials Research Letters* 10: 14. New deformation sequence of plasticity mechanisms on the nanoscopic scale.

Seymour, L., J. Maragh, P. Sabatini, M. D. Tommaso, J.C. Weaver, and A. Masic. 2023. "Hot Mixing: Mechanistic insights into the durability of ancient Roman concrete." *Science Advances* 9: Jan. 6, 2023.

Scalliet, C., B. Giuslin, and L. Berthier. 2022. "Thirty milliseconds in the life of a supercooled liquid." *Physical Review X* 12: 041028. Relaxation dynamics of model supercooled liquids simulated by molecular dynamics over a time range of 10 orders of magnitude.

Index